Dynasty of Engineers

Robert Stevenson – Founder of the Dynasty

*Painted by John Syme, R.S.A. Engraved by Thomas Dick and dedicated by him to the Northern Lighthouse Commissioners [1834]. The gold medal was presented to Stevenson in 1829 by the King of the Netherlands for the innovative distinction of a flashing revolving light (**1**, 22; **2**, 31). The original painting of c. 1833 is now at the National Portrait Gallery [PG 657].*

Dynasty of Engineers:

THE STEVENSONS AND THE BELL ROCK

ROLAND PAXTON

NORTHERN LIGHTHOUSE HERITAGE TRUST

EDINBURGH

Published by

The Northern Lighthouse Heritage Trust
84 George Street
Edinburgh
EH2 3DA

© 2011

ISBN 978-0-9567209-0-0

Printed and bound in the UK by J F Print Ltd., Sparkford, Somerset

TO THE MEMORY OF
THE LATE JEAN LESLIE (1916-2010),
ROBERT STEVENSON'S GREAT GREAT GRAND-DAUGHTER,
MY FRIEND AND SPIRITED CO-AUTHOR OF *BRIGHT LIGHTS*,
WHO WAS LOOKING FORWARD TO THE PUBLICATION
OF THIS BOOK

Illustrations

All illustrations, except where otherwise acknowledged, are from images in the author's possession. Where copyright lies elsewhere this is acknowledged under the illustration.

The background illustration for the subtitle to each Part is of a view taken from the Lighthouse Yacht, in July 1810 by artist Alexander Carse, as later redrawn by George Cumming Scott (Stevenson's apprentice from November 1817) and engraved by William Miller (**3**, *pl. XVIII*).

Contents

The Stevenson Family of Engineers

COELUM NON SOLUM

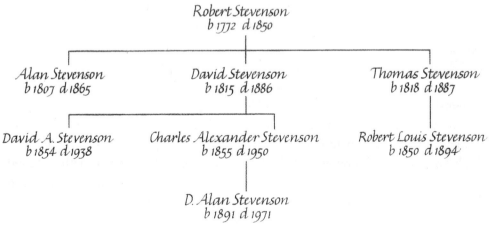

Robert Stevenson
b 1772 d 1850

Alan Stevenson
b 1807 d 1865

David Stevenson
b 1815 d 1886

Thomas Stevenson
b 1818 d 1887

David A. Stevenson
b 1854 d 1938

Charles Alexander Stevenson
b 1855 d 1950

Robert Louis Stevenson
b 1850 d 1894

D. Alan Stevenson
b 1891 d 1971

Stevenson Engineers Family Tree

For the purposes of this book,
Thomas Smith (1752–1815) – Robert Stevenson's step-father and
father-in-law – has also been included.
The dynasty was founded under Smith's patronage.

Foreword

by *Lord Boyd of Duncansby*

Chairman of the Northern Lighthouse Heritage Trust

The Northern Lighthouse Heritage Trust is pleased that this – the first book it publishes – should mark the bicentenary of the completion of the Bell Rock Lighthouse. In doing so, it celebrates the achievements of the Stevenson family of engineers. One of the main objectives of the Trust is to help promote awareness of the lighthouse heritage in Scotland and the Isle of Man, and I have no doubt that this book will achieve that.

Professor Roland Paxton's preface summarises the contents of the book. Part I is essentially factual, bringing to a wider audience the detailed biographies of the family which he wrote for the *Oxford Dictionary of National Biography* (2004), together with a fascinating insight into Robert Louis Stevenson`s experience as a reluctant engineer. Part II should be of interest to both the specialist and the general reader, as it brings to a wider audience the results of Professor Paxton`s researches into the design and building of the Bell Rock Lighthouse. Professor Paxton uses new information to prove that John Rennie – the very distinguished Scottish engineer – had a greater part in the design of the Bell Rock Lighthouse than was originally acknowledged by Robert Stevenson. During the nineteenth-century, the Stevenson and Rennie families had a number of fierce public disputes on this issue, and we hope that this book will settle the argument with recognition that both great men were responsible.

The final part of the book is a reminder that the Stevenson inheritance lives on, and provides an up-to-date list (provided by the Commissioners of Northern Lighthouses) of lighthouses that the family were responsible for in both Scotland and the Isle of Man. Virtually all of the listed lighthouses are still operating, and from their eighteenth, nineteenth, and twentieth-century structures. Although now all unmanned, twenty-first century technologies ensure that the lights still shine for, in the words from the Commissioners motto, *In salutem omnium* (i.e. the safety of all).

The Trust is enormously grateful to Professor Paxton who, as both author and editor of this book, has given freely of his time and his vast knowledge and experience of engineering and the Stevenson archives. We gratefully acknowledge the support of the OUP who allowed us to reprint Professor Paxton's biographies from the

Oxford DNB. We also thank Professor Paxton and the heirs of the late Mrs Jean Leslie, authors of *Bright Lights: the Stevenson Engineers* (1999), for allowing us to use excerpts and illustrations from this fascinating history of the Stevenson family. We are honoured too that the President of the Institution of Civil Engineers has written such a positive introduction to this celebration of the achievements of past members of the Institution.

If this book helps to remind the world of those achievements, the Trust will be content.

Lord Boyd of Duncansby
October 2010

Introduction

By *Professor Paul Jowitt*
President of the Institution of Civil Engineers

Civil engineering is the art of directing the great forces in nature for the use and convenience of man. This definition is nowhere better exemplified than in the 1807–11 erection of what is now the world's oldest continuously operational rock lighthouse on the Bell Rock, situated 11 miles off Arbroath. Its achievement by the eminent John Rennie and the relatively inexperienced Robert Stevenson (under the watchful auspices of the Northern Lighthouse Commissioners), against the seemingly impossible difficulties of an exposed site 15 feet below high water, led to a wonder of the engineering world. The project also enabled Stevenson to gain valuable experience and establish, within a decade, the private practice in which, with changes from time to time, the members of the dynasty of engineers flourished for 151 years (165 years if Smith is included). I am pleased to record that all the Stevenson participants were members of the Institution, starting from 1828.

This re-examination of the design and construction of the Bell Rock Lighthouse – based on little-known contemporary evidence – provides an authoritative account, with great images of the innovative temporary works that facilitated its erection. It also confirms Rennie's key role in the lighthouse's as-built design and overall direction, the credit for which is still generally attributed to Stevenson. The author is to be congratulated on his findings, from painstaking investigation as part of a PhD thesis (for which I was his supervisor) (**12**). He discovers that, both in terms of design and execution, this sustainable marvel of lighthouse engineering was essentially a masterpiece of joint achievement by Rennie and Stevenson in the best tradition of the chief engineer/resident engineer relationship.

The useful chronology of more than 200 lighthouses erected by the Stevensons for the Northern Lighthouse Board, enhanced by a selection of modern images, represents a legacy of which the Board and society in Scotland can be proud.

The Northern Lighthouse Heritage Trust is to be congratulated on publishing this fascinating and useful complement to the Bell Rock Lighthouse bicentenary.

Paul Jowitt
October 2010

Preface

The bicentenary of the Bell Rock Lighthouse, an engineering wonder of the world situated 11 miles off Arbroath, has created the opportunity with this book (and the generous consent of Oxford University Press), to furnish a handier version of my articles from the *Oxford Dictionary of National Biography* [*ODNB*]. These articles cover a truly remarkable dynasty of engineers.

Part I of the book comprises the biographies of the eight members of the family who, over five generations from 1786 to 1952, contributed significantly to the nation's transport infrastructure and international lighthouse engineering, particularly during the nineteenth-century. The biographies are, with minor editorial changes, as published in the *ODNB* (2004). They have been complemented with contemporary illustrations and my biographical essay on Robert Louis Stevenson's three and a half years as a reluctant trainee civil engineer, written for *Bright Lights: the Stevenson Engineers* (**2**).

Part II of the book comprises two little-known contemporary sources uncovered during my Stevenson researches which, with images and comment, shed new light on the design and erection of the Bell Rock Lighthouse and the work of its engineers. These are, following an introduction:

- A fascinating account by Robert Stevenson written in 1812 and published in 1813, which includes a fuller acknowledgement of John Rennie's part in the execution of the lighthouse undertaking than has been generally appreciated.

- A previously unpublished report by Rennie of 2 October 1809 to the Northern Lighthouse Commissioners, in which his planning and overall superintendence of the as-built lighthouse is exemplified.

Part III comprises an illustrated chronology of the Smith and Stevenson lighthouses legacy in Scotland and the Isle of Man, enhanced by Ian Cowe's fine images.

Roland Paxton

Metric equivalents

Imperial measurements have been generally adopted as this was the system used in the design and construction of the various works described.

The following are the metric equivalents of Imperial units used:

Length
1 inch = 25.4 millimetres
1 foot = 0.3048 metre
1 yard = 0.9144 metre
1 mile = 1.609 kilometres

Area
1 square inch = 645.2 square millimetres
1 square foot = 0.0929 square metre

Volume
1 gallon = 4.546 litres
1 cubic foot = 0.0283 cubic metre
1 cubic yard = 0.7646 cubic metre

Mass
1 pound = 0.4536 kilogram
1 Imperial ton = 1.016 tonnes

Acknowledgements and abbreviations

Mary Bergin-Cartwright, Birmingham City Council Archives, Ronald Birse, Lord Boyd of Duncansby, Lorna Hunter, Institution of Civil Engineers [ICE] Library, Willie Johnston, Professor Paul Jowitt, the late Jean Leslie, Peter Mackay, Roger Lockwood, Sheila Mackenzie, Virginia Mayes-Wright, Neil Miller, Alison Morrison-Low, the Trustees of the National Library of Scotland [NLS], National Archives of Scotland [NAS], National Museums of Scotland [NMS], National Portrait Gallery [NPG], Northern Lighthouse Board [NLB], Oxford University Press, Ann Paxton, Royal Commission on the Ancient and Historical Monuments of Scotland [RCAHMS], Royal Society of Edinburgh [RSE], School of the Built Environment, Heriot-Watt University, David Taylor, Whittles Publishing, the Wick Society and John Williamson.

References

[excluding those self-contained in the *Oxford DNB* text and elsewhere]

(**1**) Stevenson, A., *Biographical sketch of the late Robert Stevenson.* Edinburgh, 1861, 10: Stevenson, D., *Life of Robert Stevenson.* Edinburgh, 1878, 21–22.

(**2**) Leslie, Jean and Paxton, R., *Bright Lights: the Stevenson Engineers.* Edinburgh, 1999.

(**3**) Stevenson, R., *An Account of the Bell Rock Light-house, including the Details of the Erection and Peculiar Structure of that Edifice … By Robert Stevenson, Civil Engineer.* Edinburgh, 1824.

(**4**) Stevenson, A., *Account of the Skerryvore Lighthouse, with notes on the Illumination of Lighthouses.* Edinburgh, 1848.

(**5**) Stevenson, R. L., *Memories and Portraits.* London, 1887, 134–135.

(**6**) Engraving by W. & D. Lizars from a drawing by Stevenson's assistants W. Lorimer & J. Steedman. [In] Stevenson, R., 'Bell Rock Light-house during the gale'. *Encyclopaedia Britannica, Supplement to Fourth, Fifth and Sixth Editions.* II, 253–258. pl. Edinburgh, 1824. First published December 1816.

(**7**) Stevenson, D. A., *The World's Lighthouses Before 1820.* Oxford, 1959.

(**8**) Paxton, R., 'Account of a Visit to the Bell Rock Lighthouse on 19th–20th August 1986'. [In] 'Institution of Civil Engineers' *PHEW Newsletter*, No. 32, December 1986, 4–10.

(**9**) Ballantyne, R. M., *The Lighthouse: Being the Story of a Great Fight Between Man and the Sea.* London, 1865, facing 358.

(**10**) Paxton, R., 'ICE President unveils Robert Stevenson plaque in Edinburgh'. [In] 'Institution of Civil Engineers' *PHEW Newsletter*, No. 102, June 2004, 32, 4–10.

(**11**) Birmingham City Council Archives. MS. 3782/13/49/77. Letter: John Rennie to Matthew Boulton (Soho). 12 March 1814.

(**12**) Paxton, R., 'An assessment of aspects of the work of the Stevenson Engineers 1786–1952'. Ph.D. Thesis, Heriot-Watt University, 1999. Copies: ICE, NLS.

(**13**) Paxton, R., 'The sea versus Wick Breakwater 1863–77 – an instructive disaster'. Proc. ICE 9th Intl Coasts, Marine Structures and Breakwaters Conf. 2009.

A note on the author and editor

Professor Roland Paxton, chartered engineer and engineering historian, is a Fellow of the Institution of Civil Engineers and a Fellow of the Royal Society of Edinburgh. He was educated at Manchester College of Science and Technology and Heriot-Watt University obtaining an MSc, PhD and an honorary D.Eng. He is vice-chairman of the Institution of Civil Engineers' Panel for Historical Engineering Works (chairman 1990–2003), a trustee of the James Clerk Maxwell Foundation, chairman of the ICE Scotland Museum, and chairman of the Forth Bridges Visitor Centre Trust.

Since 1990, after an engineering career in local government, he has taught and researched in engineering history and conservation at Heriot-Watt University as an Honorary Professor, lecturing as far afield as Scandinavia, Japan, Eastern and Central USA and California. His awards include an MBE, the Institution of Civil Engineers' Garth Watson Medal and Robert Alfred Carr Prize and the American Society of Civil Engineers' History and Heritage Award. He was named Association for Preservation Technology International's *'College of Fellows Lecturer for 2000'*.

In 1996 he was instrumental in saving from collapse the world's oldest viaduct on a public railway near Kilmarnock after negotiating its purchase for £2. In 2002, with *Radar World*, he laid to rest a Highland legend by locating horse and cart remains accidentally entombed in a concrete pier of Loch-nan-Uamh Viaduct on the West Highland Railway during its construction. From 1992–2002 he served on the Royal Commission on the Ancient and Historical Monuments of Scotland.

Of more than 100 publications those most relevant to the Stevensons are, his biographical articles for the *Oxford Dictionary of National Biography*; *Bright Lights* (with the late Jean Leslie); *Civil Engineering Heritage Scotland Highlands and Islands* and *Lowlands and Borders* (with Dr Jim Shipway); and his Wick Breakwater failure findings (**13**). Other related activities have included, organising a Stevenson plaque on the Melville Column, Edinburgh, taking part in conferences and media documentaries, and serving on Bell Rock 200 anniversary committees.

Part I

Dynasty of Engineers

The Stevensons –
Biographical Articles

The following articles by the author have been reproduced with permission and minor editorial changes from the sixty-volume *Oxford Dictionary of National Biography* (2004) published by Oxford University Press:

Thomas Smith (1752–1815)
Robert Stevenson (1772–1850)
Alan Stevenson (1807–1865)
David Stevenson (1815–1886)
Thomas Stevenson (1818–1887)
David Alan Stevenson (1854–1938)
Charles Alexander Stevenson (1855–1950)
(David) Alan Stevenson (1891–1971)

with

Robert Louis Stevenson (1850–1894)
[from *Bright Lights: the Stevenson Engineers* (**2**)]

and

Chronological lists of Stevenson private firms and Northern Lighthouse Board Engineers 1787–2010

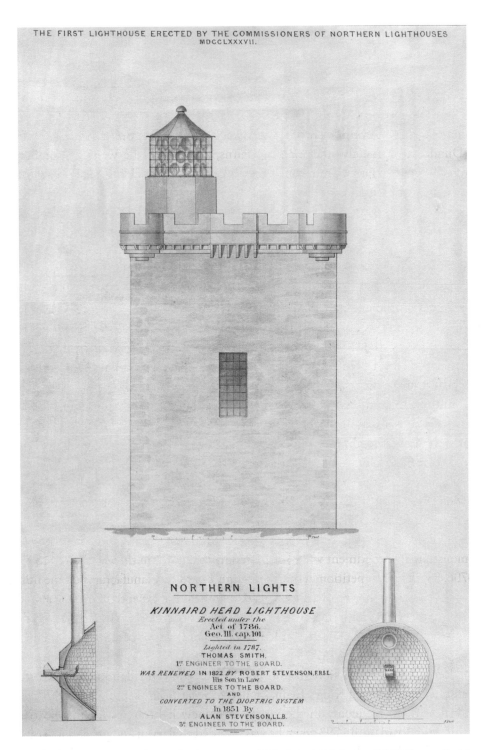

THE FIRST LIGHTHOUSE ERECTED BY THE COMMISSIONERS OF NORTHERN LIGHTHOUSES
MDCCLXXXVII.

NORTHERN LIGHTS

KINNAIRD HEAD LIGHTHOUSE
Erected under the
Act of 1786.
Geo. III. cap. 101.

Lighted in 1787.
THOMAS SMITH.
1ST ENGINEER TO THE BOARD.

WAS RENEWED IN 1822 *BY* ROBERT STEVENSON, F.R.S.E.
His Son in Law
2ND ENGINEER TO THE BOARD.
AND
CONVERTED TO THE DIOPTRIC SYSTEM
In 1851 By
ALAN STEVENSON, LL.B.
3RD ENGINEER TO THE BOARD.

*The first lighthouse erected by the Commissioners of Northern Lighthouses 1787, Thomas Smith, Engineer. [A Stevenson office drawing of 1851 or later] (**2**, 154). © NMS*

THOMAS SMITH

1752–1815

Thomas Smith (*bap.* 1752, *d.* 1815), lighting engineer, was baptized on 6 December 1752 in Ferryport-on-Craig, Fife, a small village opposite Dundee. He was the son of mariner, Thomas Smith, and his wife, Mary Kay. In 1764 Smith was apprenticed at Dundee to a metalworker named Cairns, after which he went to Edinburgh (probably in 1770), as a journeyman metalworker, when building of the 'new town' was in progress. By 1781 he was trading as a tinsmith from Bristo Street, where he manufactured oil lamps, brass fittings, fenders, and other household metal articles. His business prospered and by 1790 he had moved to premises in Blair Street, where he employed a larger workforce. He was elected to the Edinburgh Guild of Hammermen in 1789 and became its master and a city magistrate in 1802.

On 19 February 1778 Smith married Elizabeth Couper (1758–1786), daughter of a Liberton farmer. After her death, on 20 October 1787, he married Mary Jack (1762–1791). In the year following her death, on 14 November 1792, Smith married Jean Hogg, *née* Lillie (1751–1820). Jean Lillie had previously been married (in 1771), to Alan Stevenson (1752–1774) and to James Hogg, an Edinburgh gunsmith (in 1777), whom she divorced in 1792.

Smith took an interest in improving the illumination of lighthouses in 1786, before the Board of Commissioners of Northern Lighthouses was formed in the same year to improve the almost non-existent lighting of Scotland's coast. He had proposed to the Edinburgh chamber of commerce that a lamp with metallic reflectors be substituted for the coal light at the old private lighthouse on the Isle of May, but they declined their support. On 16 June 1786 Smith wrote: 'A comparative view of the supperior advantages of lamps above coal light when applyd to light houses', in which he confirmed that he had 'constructed 2 small reflectors & lamp with a view to demonstrate by experiment what has been only laid down in theory' (NLS, MS Acc. 10706, 88). He then petitioned the Edinburgh Board of Manufactures on the utility of such lamps and they resolved to allow £20 towards the expense of making a model of a reflector lamp and trying an experiment on Inchkeith: a trial that is believed to have been successful. The Northern Lighthouse Board appointed Smith as their first Engineer on 22 January 1787.

After receiving instruction in lighthouse construction and illumination in Norfolk from Ezekiel Walker of King's Lynn, Smith enthusiastically set to work in 1787. He worked, without payment, on the provision of new lighthouses for the Northern Lighthouse Board until 1793 when he was awarded a salary of £60 per annum and his expenses. The Board did not regard his lack of building and architectural experience as an impediment.

During the next two decades Smith was responsible for providing or improving 13 lighthouses, commencing in 1787 with the conversion of Kinnaird Castle into a lighthouse and followed by the Mull of Kintyre (1788), North Ronaldsay (1789), Eilean Glas (1789), and Pladda (1790) lighthouses. Independently of the Board he was responsible for harbour lights at Leith and Portpatrick, and on the Clyde and Tay rivers. His last major lighthouses were Start Point, Orkney (1802–06) and Inchkeith (1804), both for the Board, and Little Cumbrae (1793), for the Clyde Lighthouses Trust.

From 1797 Smith delegated most lighthouse matters to his apprentice and stepson Robert Stevenson, who married his daughter Jane, and established the Stevenson dynasty of engineers which practised until 1952. Stevenson formally succeeded him as Engineer to the Northern Lighthouse Board on 12 July 1808. This enabled Smith to concentrate on lamp manufacture and the expansion of his shipping and other interests, particularly his general and street lighting business. By 1800 his lamps were lighting much of eastern Scotland and the central belt as far west as Glasgow. In 1804 he was the public lighting contractor for both the Old and New towns of Edinburgh and, by 1807, for lighting the streets of Perth, Stirling, Ayr, Haddington, Aberdeen, and Leith (in 1810). In 1808 Smith retired from the business which was then carried on by his son James.

Smith developed and made arrays of parabolic reflector oil lamps. Each lamp had a light source at its focus and a curved reflector formed of small pieces of mirror glass set in plaster that produced a beam of light. His first light, at Kinnaird Head, had an intensity of about 1000 candlepower, which, although feeble compared with its modern counterpart of 690,000 candlepower, nevertheless represented a worthwhile improvement on coal lights. He retained glass-faceted reflectors for new lights until 1801, after which, because of Robert Stevenson's influence, he started to manufacture Argand lamps with silvered copper reflectors. This improvement, which produced a significantly brighter light, is believed to have been first installed in Scotland at Inchkeith Lighthouse in 1804.

Details of Smith's reflectors became more generally known from an article 'Reflector for a Light-house' in the supplement to the third edition of *Encyclopaedia Britannica* (1801). In it Smith is described as 'an ingenious and modest man [who] has carried [his inventions] to a high degree of perfection without knowing that something of the same kind had been long used in France'. This tribute was omitted from later editions, including the last carrying the article (1823), after the editor had learned of Ezekiel Walker's prior development of the glass facet reflector lamp concept. Nevertheless, Smith was the first to introduce brighter lights into Scottish lighthouses, and has a good claim to be regarded as Scotland's first lighting engineer. He died on 21 June 1815 at 1 Baxter's Place, Edinburgh, and was buried in the Old Calton Cemetery.

Sources

J. Leslie and R. Paxton, Bright Lights: the Stevenson Engineers, 1751–1971 (1999) · C. Mair, Star for Seamen: the Stevenson Family of Engineers (1978) · D. A. Stevenson, The World's Lighthouses Before 1820 (1959) · Edinburgh Advertiser (30 June 1815) · Edinburgh Evening Courant (1 July 1815) · Caledonian Mercury (1 July 1815) · 'Reflector for a Light-house', Encyclopaedia Britannica, 3rd edn, suppl. (1801) · R. L. Stevenson, Records of a Family of Engineers (1912) · Private information (2004)

Archives

Edinburgh City Archives, Chamber of Commerce at Edinburgh MSS · NA Scot., Board of Manufactures and Carron Co. MSS · NA Scot., Northern Lighthouse Board minute books · NLS., Business Records of Robert Stevenson & Sons, Civil Engineers, MS Acc. 10706

Wealth at death

Approx. £26,000: Private information

*Inchkeith Lighthouse tablet 1804 (**2**, 21).*

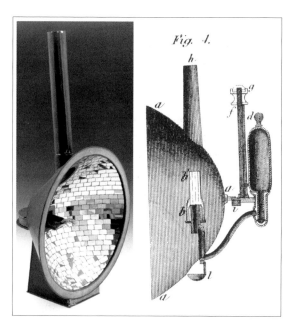

*Left: Model of Smith's multi-facetted glass reflector lamp [National Museums of Scotland ref. 1868. L19. Neg. 13744] © National Museums of Scotland with a section of an early nineteenth-century continuous silvered-copper reflector lamp with Argand burner and wick. (**3**, pl. XX)*

ROBERT STEVENSON
1772–1850

Robert Stevenson (1772–1850), civil engineer, was born in Glasgow on 8 June 1772. He was the only child of Alan Stevenson (1752–1774), a West India merchant, and his wife, Jean Lillie (1751–1820). Two years later his father died of fever in St Kitts, leaving his family in straitened circumstances, and Stevenson was educated at a charity school in Edinburgh. In 1786 he was apprenticed to an Edinburgh gunsmith and was himself described as a gunsmith about 1791. At about that time he began work for Thomas Smith (bap. 1752, d. 1815), an Edinburgh tinsmith, lampmaker, and merchant who, following his invention of a light reflector, had been appointed Engineer to the newly formed Northern Lighthouse Board in 1787. Smith married Stevenson's mother in 1792 and became his father-in-law when Stevenson married his daughter Jane (*c*.1779–1846) on 3 June 1799.

During the winters of 1792–94 Stevenson attended Professor John Anderson's classes in natural philosophy at Glasgow University and was directed by him towards an engineering career. From 1796 until 1802 he was apprenticed to Smith, specializing in lighthouse work, and gained experience on reflector installation, building maintenance, and construction of the Pentland Skerries and Little Cumbrae lighthouses. From 1797 Stevenson exercised considerable autonomy in the firm's lighthouse work including the construction of Inchkeith and Start Point lighthouses. By 1802 he had been taken into partnership by Smith, whom he succeeded as Engineer to the Board in 1808. During the winters of 1800–04 Stevenson continued to develop an engineering career by attending classes at Edinburgh University in mathematics, natural philosophy, chemistry, and natural history. During this period he was also trying to gain approval for the Bell Rock Lighthouse project 11 miles off Arbroath, which was to prove his most important engineering achievement.

In 1799, following a storm in which many ships were wrecked, Stevenson had proposed erecting on the Bell Rock a beacon-style lighthouse on cast-iron pillars. In 1800 however, after seeing that the rock was submerged by about 12 feet at each high tide, and considering the possibility of damage by ships, he abandoned this idea in favour of a more substantial lighthouse to be made of stone. As part of the design and promotional process for the project both designs were accurately modelled, a practice which Stevenson often employed subsequently on important work. Because of the hazardous and expensive nature of the project it was only after the Board had obtained the support of the eminent engineer John Rennie in 1805 that the necessary act of parliament was passed in 1806. Rennie was appointed Chief Engineer and with Stevenson as Resident Engineer the lighthouse was constructed in 1807–11.

Rather than in its design, the great achievement of the work was in the exceptional difficulty of its execution, which was carried out by Stevenson and his dedicated workmen. The design at Rennie's insistence was more closely modelled on Smeaton's Eddystone Lighthouse than Stevenson's proposal, and this was particularly evident in respect of its external shape. Stevenson also had been strongly influenced by the Eddystone design and improved on it in detail with cantilevered and bonded, instead of flat-arched floors to compartments – an innovation [as approved and developed by Rennie – see page 77] adopted in subsequent rock lighthouses. Rennie recognized the importance of Stevenson's role when he wrote to him in 1807 that the work 'will if successful, immortalise you in the annals of fame' (Stevenson, Biographical Sketch, 10). Innovations introduced under Stevenson's direction included the temporary beacon barrack, elevated cast-iron railways across the rock, and the ingenious movable jib and iron balance cranes, (records indicate these were invented by foreman millwright Francis Watt who also designed the barrack as built). The success of the work enabled Stevenson, from 1811, to establish within a decade one of Scotland's leading indigenous civil engineering practices. With his descendants, he practised engineering continuously until 1952. His classic *Account of the Bell Rock Light-house* was published in an edition of 300 copies in 1824 and he can be considered to have attained the first rank of his profession shortly thereafter. The lighthouse is still in service, but unmanned.

As Engineer and Chief Executive to the Northern Lighthouse Board in 1808–43, Stevenson can be said to have inaugurated the modern lighthouse service in Scotland. He designed and constructed at least 18 lighthouses including Toward Point (1812), Isle of May (1816), Corsewall (1817), Point of Ayre and Calf of Man (1818), Sumburgh Head (1821), Rhinns of Islay (1825), Buchan Ness (1827), Cape Wrath (1828), Tarbat Ness and Mull of Galloway (1830), Dunnet Head (1831), Douglas Head (1832), and Girdle Ness, Barra Head, and Lismore in 1833. Stevenson developed Smith's work on lighthouse illumination and brought the catoptric system, using silvered copper parabolic reflectors and Argand lamps, to a high degree of perfection. With the increasing number of lights it became necessary to distinguish between them, and to this end he devised intermittent and flashing lights.

Marine engineering represented the largest element of Stevenson's general practice. He proposed improvements at numerous harbours including Dundee, Peterhead, Stonehaven, Sunderland, Fraserburgh, and Granton, and river navigation schemes for the Forth, Tay, Severn, Mersey, Dee, Ribble, Wear, Tees, and Erne. Many of these proposals were implemented. He also reported on ferry crossings of the Forth, Tay, Dornoch and Pentland firths, and the Severn, and on fisheries. Stevenson's state-of-the-art marine work included the design and construction in 1821 of a sea wall with a cycloidal-curve vertical profile which dissipated wave energy more effectively than common walls. His experiments on the destruction of timber by the *Limnoria terebrans* influenced the universal adoption of greenheart for marine timberwork. In Aberdeen in 1812, he discovered that salt water from the ocean flowed up river in a distinct layer from the fresh water which overflowed it. This led to his invention of the 'hydrophore' or water sampler for procuring samples, to further experiments, and to his Royal Society of Edinburgh paper 'On Vertical Differences of Salinity in Water' (1817). His 'Observations upon the Floor of the North Sea' delivered to the Wernerian Society, were published in 1817 and 1820. Stevenson was joined in partnership of the firm by his sons Alan Stevenson (in about 1832) and David Stevenson (1838). In 1846, Thomas Stevenson became a partner on his father's retirement.

From 1811 to 1827 Stevenson was extensively engaged on canal, road, and railway projects, often adopting a promotional role. Before 1818 he made proposals for canals on one level between Edinburgh and Glasgow and also in the Vale of Strathmore. In 1828 he worked with Telford and Nimmo on a proposal for a new harbour at Wallasey and a ship canal across the Wirral to the Mersey. None of these schemes were executed but he was more successful with road making and in his advocacy of stone tracks in city roads.

By 1818 Stevenson was convinced of the superiority of railways over small canals for inland communication and proposed the Edinburgh Railway to connect with the Midlothian coalfield. In 1819 he advised on the line for the Stockton and Darlington Railway. By 1820 he was the leading authority on horse-traction railways in Scotland and he edited, with notes, the numerous 'Essays on Rail-roads' submitted to the Highland Society, which were published in 1824. By 1836 he had worked out various railway schemes to traverse eastern Scotland from the Tweed to Perth and Aberdeen, and from Edinburgh to Glasgow via Bathgate. These were more or less on the lines of the eventual railway network, but the financial climate was unfavourable and the necessary finance for their implementation was not forthcoming. The only scheme actually constructed was the short Newton Colliery Railway to Little France near Edinburgh. Stevenson's design practice was basically the same as he had adopted for canals: to plan his railways as near level as practicable, using stationary steam-engine powered inclined planes to overcome differences in level. In 1818 he advocated the use of 12 feet long malleable iron edge-rails in preference to the much shorter and weaker cast iron rails then prevalent. Three years later George Stephenson acknowledged Stevenson's influence on Birkinshaw's epoch-making development (the malleable iron forerunner of the modern steel rail) and wrote to him 'you have been at more trouble than any man I know of in searching into the utility of railways' (Stevenson, Biographical Sketch, 27).

Stevenson was also a notable bridge engineer and highway planner. He designed and constructed many bridges throughout Scotland including, over the Clyde at

Glasgow, Hutcheson Bridge (1832–68) and a temporary but notably wide fourteen-span timber bridge (1832–46). The former, which had to be replaced because of Clyde navigation deepening, was considered one of the best specimens with segmental masonry arches. So was Stirling Bridge which still stands, Stevenson also planned its town approach. This approach is not as imposing as his earlier London and Regent Road approaches into Edinburgh. These skirt Calton Hill, and include the Regent's Bridge with its open parapets, which enable users to enjoy the view. Stevenson was also responsible for making these roads, which involved blasting, rock excavation, and building a massive retaining wall.

Segmental arches characterize Stevenson's masonry bridges, fine examples of which still exist at Marykirk, built in 1812, and Annan, in 1824. He adopted segmental arches in major proposals for cast iron additions to Newcastle upon Tyne, Perth, and Edinburgh North bridges, but none was executed. Stevenson's innovative designs for other types of bridges included a laminated timber arch for Dornoch Firth in 1830. From 1821, he proposed a new type of medium-span suspension bridge without towers for numerous locations. This design was novel in that the roadway superstructure rested on the catenarian chains rather than being suspended from them. It was widely publicized in Stevenson's authoritative 'Bridges of Suspension' article published in the Edinburgh Philosophical Journal (1821), and translated into French, German, and Polish. By 1850, numerous small-span bridges on this basic principle were subsequently executed in Britain and on the continent. Although not implemented as intended, Stevenson's proposals, together with details of his Glasgow and Stirling bridges, were widely disseminated through John Weale's *The Theory, Practice and Architecture of Bridges* (1839). They undoubtedly influenced subsequent bridge-building practice nationally and more unusual structures upon which Stevenson advised, were the cracked steeple of Montrose church, Arbroath Abbey, and the Melville Monument in Edinburgh.

Stevenson had a lifelong interest in gaining and promoting knowledge and his writings appeared in more than 60 publications. Many were engineering reports, but about one-third achieved much wider circulation through leading periodicals and encyclopaedias. He contributed significant articles to the *Edinburgh Encyclopaedia* between 1810 and 1824, entitled 'Bell Rock', 'Eddystone Rocks', 'Inchkeith', 'Light house', 'Roads and Highways', and 'Railway'. He contributed 'Bell Rock Light house', 'Blasting', 'Caledonian Canal', and 'Dredging' to the *Encyclopaedia Britannica* between 1816 and 1819.

In 1817 he wrote a lively and informative series of letters to his sixteen-year-old daughter, Jane, while on a tour through the Netherlands; these were of sufficient interest to be published in the *Scots Magazine* (1818–21) and separately as *Journal of a Trip to Holland* (1848). Many of Stevenson's publications, because of their depth and authority, now represent a valuable historical resource. His membership of learned societies seems to have commenced with that of the Highland Society in 1807. By 1812 he was a member of the council of the Wernerian Natural History Society and in the following year a founder director of the Astronomical Institution of Edinburgh. In 1815 he was elected to fellowships of the Royal Society of Edinburgh, the Geological Society, and the Society of Antiquaries of Scotland. Six years later he became a founder member of the Scottish Society of Arts and in 1827–28 was elected to membership of the Smeatonian Society of Civil Engineers and the Institution of Civil Engineers.

Prominent points of Stevenson's character, noted by his sons, were his sagacity, fortitude, perseverance, unselfishness, generosity, a high sense of duty, and his extensive and unwearied exertions in forwarding the progress of young professional men. He was a member of the Church of Scotland and an elder, first at St Mary's, Edinburgh, from 1828–43 and afterwards at Greenside parish church. He died at his home, 1 Baxter's Place, Edinburgh, on 12 July 1850 and was buried in the New Calton Cemetery. An affectionate portrait is given by his grandson Robert Louis Stevenson (Stevenson, *Records*).

Sources

D. Stevenson, Life of Robert Stevenson (1878) · A. Stevenson, Biographical Sketch of the Late Robert Stevenson (1861) · C. Mair, Star for Seamen: the Stevenson Family of Engineers (1978) · R. Stevenson, Report Relative to Various Lines of Railway from the Coal-field of Midlothian to … Edinburgh (1818) · D. Brewster and others, eds., The Edinburgh Encyclopaedia, 3rd edn, 18 vols. (1830) · Encyclopaedia Britannica, suppl. to 4th–6th edns (1824) · R. Stevenson, 'Vertical Differences of Salinity in Water', Annals of Philosophy, 10 (1817), 55–58 · Business Records of Robert Stevenson & Sons, NLS, Acc. 10706 · R. L. Stevenson, Records of a Family of Engineers (1912) · Private information (2004) · Family Grave, New Calton Cemetery

Archives

NLS, Business Records of Robert Stevenson & Sons, Acc. 10706

Likenesses

J. Syme, oils, c.1833, Scot. NPG [see illus.] · T. Dick, engraving, c.1840 [1834?] (after J. Syme), priv. coll. · S. Joseph, marble and bronze (?) bust, Northern Lighthouse Board, 84 George Street, Edinburgh · S. Joseph, plaster bust, Scot. NPG · Plaster bust (after S. Joseph), Scot. NPG

Wealth at death

£15,154 13s. 5d.: NAS.

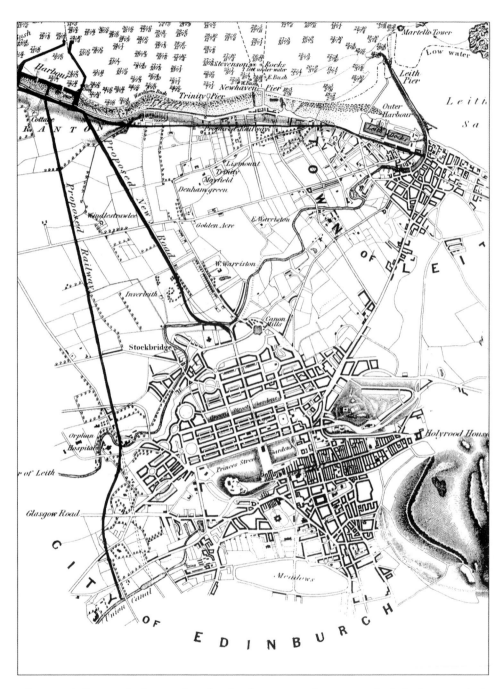

Stevenson's Granton harbour and road and railway proposals in 1834, of which about 800 ft of pier, and Granton Road, were built. From 1815-19 he also engineered the picturesque Regent Road approach to the East End of Princes St., Edinburgh, skirting Calton Hill and including the Regent Arch. From Stevenson & Son's Granton Harbour Report, 1834. (2, 57)

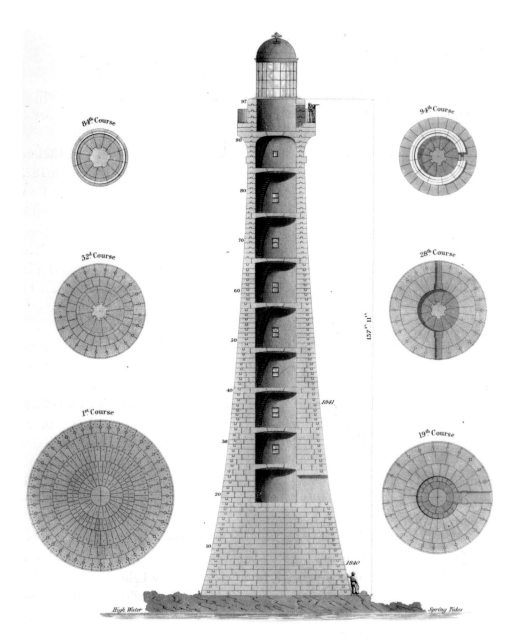

Alan Stevenson's Skerryvore Lighthouse – one of the world's most finely proportioned, dubbed by R. L. Stevenson the 'noblest of all deep sea lights' (5). Tower 155ft high, weight c.4100 tons, with hyperbolic side curvature. Note: Foundation above High Water – lower courses undovetailed.
(4, pl. VIII)

ALAN STEVENSON
1807–1865

Alan Stevenson (1807–1865), civil engineer, was born on 28 April 1807 in Edinburgh. He was the eldest surviving son of Robert Stevenson, the brother of David Stevenson and Thomas Stevenson and the uncle of Robert Louis Stevenson. Educated at the High School and University of Edinburgh, where from 1821 he read Latin, Greek, and mathematics with a view to entering the church. In 1823 Stevenson decided to follow an engineering career and, after six months at the Revd Pettingal's Twickenham Finishing School, he commenced a four-year apprenticeship to his father (a civil engineer). During each winter he attended classes at Edinburgh University appropriate to his career and in 1826 he graduated MA. Under Sir John Leslie (1766–1832), professor of natural philosophy, he gained the Fellowes prize for excellency as an advanced student of natural philosophy. In the same year he read his first paper to the Royal Physical Society of Edinburgh, on the 'causes of obscurity in style', urging for greater clarity in writing.

Stevenson's thorough training from 1823–27 included railway surveying under William Blackadder of Glamis; bridge building, river improvements, harbour and lighthouse engineering under his father's direction; and a study tour of works in Sweden and Russia with the engineer Robert Bald. In 1828–29 he gained experience on works conducted by Thomas Telford and James Walker at Hull docks, on the Birmingham Canal under its Resident Engineer, William Mackenzie, and on a new dock at Liverpool. Encouraged by his father, and with access to his notes, Stevenson compiled a list of British lighthouses, describing their appearance at night. This pocket-book list, the first of its kind, published at Leith in 1828 as *The British Pharos* (2nd edn, 1831), was very useful to mariners.

In 1830 Stevenson, in addition to being an assistant in his father's firm, was appointed clerk of works to the Northern Lighthouse Board, to which his father was Engineer. During the following three years he worked on new lighthouses at the Mull of Galloway, Dunnet Head, Douglas Head, Barra Head, Girdleness, and Lismore. For some years Stevenson had been interested in improving lighthouse illumination and in the summer of 1834 he visited lighthouses and workshops in France. There he gained knowledge of the work of the Fresnel brothers – Augustin (1788–1827) and Léonor (1790–1869), Jean Baptiste François Soleil (1798–1878) and Bordier-Marcet Issac-Ami, (1768–1835). In particular, he studied Fresnel's dioptric apparatus which used lenses instead of mirror reflectors to enhance light intensity.

In the following year Stevenson's influential Report 'On Illumination of Lighthouses by Means of Lenses', which included a valuable account of French practice, was published. In 1835 under his direction, the revolving light at Inchkeith and in

1836 the fixed light of the Isle of May were made dioptric instead of catoptric, with a resulting threefold order of increase in brightness. Furthermore, for Trinity House in 1836 Stevenson designed and superintended the installation of the first dioptric light in England at Start Point, Devon.

In around 1832 Stevenson was taken into partnership in his father's firm. From 1832 to 1837, he and his father were engaged in work which included, preparation of 'A Chart of the Coast of Scotland', Ballyshannon harbour improvement, Granton harbour, plans for Edinburgh and Glasgow, Edinburgh and Dundee, and Perth and Dunkeld railways, Perth harbour, and Tay and Ribble navigation improvements. Soon afterwards he wrote an authoritative account of 'sea lights' from their earliest development to about 1838 which was published in the seventh and eighth editions of the *Encyclopaedia Britannica* (1840 and 1857). From December 1837 until August 1843, Stevenson was almost exclusively employed by the Northern Lighthouse Board on the design and construction of Skerryvore Lighthouse. This being mutually agreed as too arduous a task for his father who was by then sixty-five.

In January 1843 Stevenson succeeded his father as the Board's Engineer. From then, until paralysis dictated his retirement in 1853 he was responsible for the design and construction of new lighthouses. In addition to Skerryvore, these included Little Ross, Covesea Skerries, Chanonry Point, Cromarty, Cairn Point (Loch Ryan), Noss Head, Ardnamurchan, Sanda, Heston Island, Hoy, and Arnish Point, Stornoway. On 11 September 1844 he married Margaret Scott (1813–1895), daughter of Humphrey H. Jones of Llynon, Anglesey and his wife, Jean, *née* Scott. They had one son – the art critic Robert Alan Mowbray Stevenson (1847–1900) – and three daughters including the author Katharine de Mattos.

Stevenson's national reputation was mainly based on his design and execution of Skerryvore Lighthouse. His classic account, was, together with his notes on lighthouse illumination, published in an *Account of the Skerryvore Lighthouse* (1848). These notes were extended and more widely propagated through *A Rudimentary Treatise on the History, Construction and Illumination of Lighthouses* (1850). Both works were considered of outstanding technical value well into the twentieth-century. Skerryvore Lighthouse was a great engineering achievement and is still in service. It stands 155ft high, located on an isolated reef 12 miles west-south-west of Tiree and exposed to the full fetch of the Atlantic.

The first season's work, the beacon-barrack erected in 1838, was totally destroyed by a November storm. The eventual creation of the lighthouse by 1843 severely tried Stevenson's courage, patience, and health and fully exercised his undoubted ability. The lighthouse has been widely acknowledged to be the finest example for mass combined with elegance of outline of any rock tower. Stevenson adopted for its shape the hyperbolic curve which was the form with the least mass and lowest centre of gravity of the various options examined. It strongly influenced the design of the Alguada Lighthouse built for the Indian government in 1862–65. Skerryvore's revolving dioptric apparatus was the most advanced in the world at that time, with prismatic rings instead of mirrors below the central belt, thus greatly extending the improved dioptric effect. Stevenson further improved its efficiency by introducing inclined astragals into the lantern. His improvements to the dioptric system, which included the conversion of Fresnel's narrow lenses in fixed systems into a truly cylindrical drum, led to its wider adoption. He introduced prismatic rings above and below the central belt, thus securing equal distribution of light all round and extending dioptric action through the whole height of the apparatus.

In 1830 Stevenson, sponsored by Telford, became a corresponding member of the Institution of Civil Engineers, and in 1838 a fellow of the Royal Society of Edinburgh, acting as a member of its council in 1843–45. In 1840 the University of Glasgow conferred on him the honorary degree of LL.B. The emperor of Russia and the kings of Prussia and the Netherlands presented him with medals in acknowledgement of his merit as a lighthouse engineer. Dr John Brown wrote that Stevenson had genuine literary genius, that he was able to read Italian and Spanish critically and with ease, and that he knew Homer by heart and read Aristophanes in Greek. Clarity and style characterize his writings. In addition to the publications already mentioned, Stevenson contributed papers to the *Edinburgh New Philosophical Journal* and in 1851 his biographical sketch of his father was published (illustrated edn, 1861).

In 1852 Stevenson was seized with paralysis, and in the following year he resigned as Engineer to the Northern Lighthouse Board. He beguiled his suffering in retirement by translating the ten hymns of Synesius, bishop of Cyrene AD 410. These translations, along with other poems, were printed for private circulation in 1865. On 23 December 1865 Stevenson died of general paralysis at his home, 6 Pitt Street, Portobello and was buried in the New Calton Cemetery, Edinburgh. He was survived by his wife and their four children. On 3 January 1866 the Northern Lighthouse Board recorded their deep and abiding regrets for the loss of a man whose services had been to them invaluable and whose works combined profound science and practical skill.

Sources

The Scotsman (26 Dec 1865) · Proceedings of the Royal Society of Edinburgh, 6 (1866–69), 23–5 · PICE, 26 (1866–67), 575–7 · C. Mair, Star for Seamen: the Stevenson Family of Engineers (1978) · Business records of Robert Stevenson & Sons, NLS, Acc. 10706 · Private information (2004) · DNB · d. cert.

Archives

NLS, business records of Robert Stevenson & Sons, Acc. 10706.

Likenesses

Photograph, repro. [in] Mair, Star [better in family group overleaf]

Wealth at death

£9,388 16s. 4d.

The Stevenson family at Anchor House, North Berwick c. 1860.
David (on right) Alan (centre-middle) with David Alan on his knee and Charles (on his
right with stick) Thomas (back-left) and Robert Louis (front-left, 2nd from end). (2, 161)

DAVID STEVENSON

1815–1886

David Stevenson (1815–1886), civil engineer, was born in Edinburgh on 11 January 1815. He was the third surviving son of Robert Stevenson and the brother of Alan Stevenson and Thomas Stevenson. He was educated in Edinburgh at the High School in 1824–30 and attended classes at the University in 1831–35 under professors Wallace, Forbes, Hope, and Jameson while undergoing an exemplary engineering training in his father's office. Before commencing his apprenticeship Stevenson gained experience in working with iron and wood with a leading Scottish millwright, James Scott of Cupar, Fife. He was also taught mechanical drawing by James Scott's son, David. In 1832 Stevenson gained experience on bridge works being conducted by his father over the Clyde at Glasgow, and the Forth at Stirling where he dressed at least one arch stone. He gained further experience at the cotton works of James Smith of Deanston (1789–1850), agricultural engineer. In 1833–34 Stevenson's training included surveying for the Tay navigation improvement and with contractor William Mackenzie (1794–1851), setting out and starting work on a 22-mile section of the London and Birmingham Railway in Warwickshire and on the construction of Edgehill Tunnel, Liverpool.

In 1835 Stevenson studied road making in Ireland and diving bell work at Kingstown harbour. He also made a survey of, and supervised, road construction on the Calf of Man. His earliest papers were read to the Royal Scottish Society of Arts on the subjects of the Liverpool and Manchester (1835) and Dublin and Kingstown railways (1836). The former earned him a medal from the society. Late in 1835, aged twenty, Stevenson declined an invitation from Marc Isambard Brunel to work as a resident engineer on the Thames Tunnel, preferring the post of Resident Engineer at Granton harbour. Here he worked under his father on quarry opening, pier construction, and road building. In 1837 he resigned this post and gained wider experience in North America, France, Switzerland, Germany, and the Netherlands. This tour led to the publication of his influential and now historically valuable *Sketch of the Civil Engineering of North America* (1838, 2nd edn, 1859), which influenced the introduction into Britain of faster steam vessels, with fine lines and long-stroke pillar engines and cost-effective timber construction.

In May 1838 Stevenson entered into partnership with his father and brother, Alan, and the firm became known as Robert Stevenson & Sons. As 'managing partner' he immediately became responsible for the entire management of the firm's general business. Work in Scotland included navigational improvement on the Forth, Tay, Clyde, and Nith, and harbour construction. English projects included improvements to the Dee, Lune, Ribble, Wear, and Fossdyke, and in Ireland, the Erne and Foyle.

Other work included Mullagmore and Morecambe harbours, Allanton Bridge, Newfoundland lighthouses, Peebles railway, Birkenhead docks, and opposing railway crossings of the Tay and Dee. On 3 June 1840 Stevenson married Elizabeth (1816–1871), daughter of James Mackay, a goldsmith from Edinburgh. They had four sons and four daughters. Only two sons survived childhood; they were David Alan Stevenson (1854–1938) and Charles Alexander Stevenson (1855–1950) [*see under Stevenson, David Alan*], who continued the family engineering tradition.

Stevenson also advised on salmon fishing disputes and in 1842, in connection with the Dornoch fishings, first categorized the different physical characteristics of a river in to the well-known terms 'sea proper', 'tidal', or 'river proper'. This work led to his paper 'Remarks on the Improvement of Tidal Rivers' read to the Royal Society of Edinburgh in 1845, which was also separately published, under the same title, that year (2nd edn, 1849). In it he argued conclusively that if the duration of tidal influence was extended, the hydraulic head would be lessened and the velocity of tidal currents decreased. Stevenson also correctly propounded the theory of the origin of bars at the mouths of rivers and defined effective measures for their removal. He emphasized the necessity for accurate data upon which to base improvements and wrote *A Treatise on the Application of Marine Surveying and Hydrometry to the Practice of Civil Engineering* (1842). Stevenson's practice of marine engineering was extensively promoted through his article 'Inland Navigation' in the *Encyclopaedia Britannica* (8th edn, 1857) and enlarged into *Canal and River Engineering* (1858). This became a definitive work which continued to be used well into the twentieth-century and established his national reputation.

In 1846, when railway projects (and to a lesser extent public health bills), were overwhelming the Admiralty and Woods and Forests departments, Stevenson held courts of inquiry under the Preliminary Inquiries Act for at least 20 proposed bills. His findings were accepted in every case, except for the Caledonian Railway's proposed crossing of the Clyde at Glasgow. This was at first opposed by the Admiralty but later approved. In 1849–50 he reported on Fishery Board proposals at Lybster and Scallisaig harbours, which led in 1851 to his becoming, at his own request, joint Engineer to the Board with his brother Thomas for over 30 years. In 1853 Stevenson succeeded his brother Alan as Engineer to the Northern Lighthouse Board. During the following year he achieved, under difficult circumstances, the construction of Britain's most northerly lighthouse at North Unst (Muckle Flugga). In 1855, again at his own instigation, he became joint Engineer to the Board with Thomas. In 1855–80 they designed and executed some 28 beacons and 30 lighthouses, two of which, Dhu Heartach (1872) and Chicken Rock (Isle of Man) (1875) were works of particular difficulty on isolated rocks.

The general business of what had become the firm of D. and T. Stevenson continued to flourish under Stevenson's management until his retirement due to ill health in 1884. In addition to marine work the firm was also engaged on public health improvements, including Edinburgh and Leith sewerage in 1863 – the city's first such major scheme – which involved the construction of the Water of Leith sewer to an outfall in the Forth. Abroad, the firm's lighthouse practice extended to India, New Zealand, and Japan, including the organization of complete systems for the two latter countries.

For Japanese lighthouses in earthquake zones Stevenson devised an 'aseismatic' joint to mitigate the effect of shocks on lighting apparatus. However, according to Richard Henry Brunton (1841–1901), the site engineer the firm had recruited, it did not prove effective in practice. Stevenson played a leading part in developing and promoting the use of paraffin in place of the more expensive colza oil in lighthouse illumination (from around 1870). This resulted in enhanced light intensity at a lower cost and considerable savings worldwide.

In 1844 Stevenson was elected a fellow of the Royal Society of Edinburgh (vice-president in 1873–77), and also a member of the Institution of Civil Engineers; he contributed papers to both bodies. Stevenson was also a member of the Société des Ingénieurs Civils, Paris, and other learned societies. He was twice president of the Royal Scottish Society of Arts (in 1854 and 1869), his latter presidential address being entitled 'Altered relations of British and Foreign Industries and Manufactures'. He was also Engineer to the Convention of Royal Burghs of Scotland and to the Highland and Agricultural Society. Other writings by Stevenson included 'Our Lighthouses' in *Good Words* (1864) (which was also separately published in book form in the same year), *Reclamation and Protection of Agricultural Land* (1874), and the definitive life of his father published in 1878. His many interests included the improvement of agricultural implements, better endowment of professorial chairs of the University of Edinburgh, art and art criticism, and the formation of a valuable collection of etchings and engravings.

Stevenson was a man of sound judgement, upright, kind, open, and easily accessible. He died at Anchor Villa, West Links, North Berwick, of apoplexy on 17 July 1886 and was buried at Dean Cemetery, Edinburgh.

Sources

Proceedings of the Royal Society of Edinburgh, 14 (1886–7), 145–51 · PICE, 87 (1886–7), 440–43 · C. Mair, Star for Seamen: the Stevenson Family of Engineers (1978) · J. K. Finch, Engineering Classics of James Kip Finch (1978) · D. Stevenson, The Principles and Practice of Canal and River Engineering, rev. D. A. Stevenson and C. A. Stevenson, 3rd edn (1886) · Business records of Robert Stevenson & Sons, NLS, Acc. 10706 · Private information (2004) · d. cert.

Archives

NLS, business records of Robert Stevenson & Sons, Acc. 10706

Likenesses

Engraving, repro. [in] Mair, Craig, Star for Seamen; (2, 163)

Wealth at death

£55,025 13s. 11d.

Western Water steam-boat on River Ohio, 1837. From David Stevenson's drawing in his Sketch of the Civil Engineering of North America, *1838. [Pl. V].*

*Thomas Stevenson © The Royal Society of Edinburgh (**2**, 83)*

THOMAS STEVENSON

1818–1887

Thomas Stevenson (1818–1887), civil engineer and meteorologist, was born in Edinburgh on 22 July 1818. He was the youngest surviving son of Robert Stevenson and his wife, Jane, and the brother of Alan Stevenson and David Stevenson. Educated in Edinburgh at Alexander Brown's Preparatory School and at the High School, his performance was unremarkable, except for acquiring a grounding in Latin, which he cultivated and enjoyed in later years. Brown's contagious enthusiasm for English literature found in Stevenson a receptive mind, but the severe discipline which accompanied it initiated his lifelong contempt for formal education. On leaving school he provisionally entered the printing firm of his father's friend Patrick Neill, but he did not pursue a typographical career. Youthful interests included collecting books, writing, and printing some of his essays on his own working model of a 'Columbian' press.

At the age of 17 Stevenson entered the family engineering firm and during a rigorous apprenticeship to his father in 1836–39 he gained experience in harbour, river improvement, and lighthouse work and attended several classes at Edinburgh University. He still found time to write fiction, amounting to a 'drawer full', but when this was discovered by his father he was urged by letter to 'give up such nonsense and mind your business' (Mair, 143). By 1841 Stevenson's book collection is known to have embraced Aesop's *Fables* with Bewick woodcuts, Boece's *Croniclis of Scotland* (1527), and engineering-related works, including Sinclair's *Hydrostaticks* (1672) with skilful pen and ink restoration of missing plates, presumably carried out by its young owner.

From 1839 until 1841 Stevenson combined his talent for writing with engineering in the columns of *The Civil Engineer* and *Architect's Journal*. His annotated copies attest to communications advocating the removal of ruinous buildings by blasting, and restoration rather than replacement of ruinous historic buildings; on the form of river bank profiles; the repair of breaches; and on an improved levelling staff and 'quick-set' level, made to his design. In 1845 he furnished the *Journal* with an abstract of his paper to the Royal Society of Edinburgh on forces exerted by sea-waves, the first significant work on this subject. These contributions, together with the first of numerous articles in the *Edinburgh New Philosophical Journal* (in 1842–43) on defects in rain gauges and the geology of Little Ross Island, were the earliest of more than 60 publications during his lifetime. Many of these demonstrate Stevenson's innate faculty for the quantitative investigation of natural phenomena and artificial constructions which enabled him to advance contemporary knowledge and practice by means of observation and experiment.

On his father's retirement in 1846 Stevenson became the junior partner in the firm, which after a short period as Messrs Stevenson soon became known as D. and

T. Stevenson. In this capacity, he was able for many years to engage effectively in research and development and became a leading authority on lighthouse illumination and harbour engineering. Abroad, the firm's advice extended to lighthouses in India, China, and Newfoundland and to the lighting of the whole coasts of Japan and New Zealand. In Scotland in 1855–84, Stevenson acted jointly with his brother David as Engineer to the Northern Lighthouse Board; they designed and executed some 28 beacons and 30 lighthouses, including Dhu Heartach (completed in 1872) and Chicken Rock (Isle of Man) (completed in 1875), on isolated rocks. He then acted as sole Engineer to the board for nine months and as joint Engineer with his nephew David Alan in 1885–57 until his death.

From 1851 for several decades the firm acted as engineers to the British Fisheries Society and Fishery Board working at Lybster, Wick, Peterhead, and other harbours. Stevenson continued his practical investigations into the generation and force of waves. By 1852 he had formulated a tentative empirical relationship between their height and fetch which was commonly used by engineers for more than a century afterwards as a first approximation. Other experiments led to formulae which enabled the effect of harbours and breakwaters in reducing the height of waves to be calculated. Stevenson's valuable work became widely known through his *Encyclopaedia Britannica* article 'Harbour' (1857) separately published as *The Design and Construction of Harbours* (1864; 2nd edn, 1874; 3rd edn, 1886). The firm's harbour work was almost invariably successful, except for Wick breakwater which, as its construction progressed, proved unable to resist the effect of storm-driven waves in 1872–73 and was eventually abandoned as a costly, but nevertheless most instructive, failure [13].

Stevenson's national reputation was based on his harbour work and more particularly on his devices, by which 'the great sea lights in every quarter of the world now shine more brightly' (R. L. Stevenson, *Familiar Studies*, v). These are fully described in his classic work *Lighthouse Illumination* (1859; 2nd edn, 1871), expanded into *Lighthouse Construction and Illumination* (1881). He developed the work of Augustin Fresnel and Alan Stevenson and installed at Peterhead North Harbour Lighthouse in 1849 a catadioptric [reflector and lens-prisms] fixed holophote, that is, a device which was the first to combine the whole sphere of rays diverging from a light source into a single beam of parallel rays. Stevenson then further developed this system by introducing the first dioptric [lens-prisms] holophotal revolving light which was installed at Horsburgh Rock near Singapore in 1850.

This holophotal system, which proved a great improvement in lighthouse illumination, was then adopted on a larger scale by the Northern Lighthouse Board at North Ronaldsay Lighthouse in 1851 and afterwards came into universal use. Stevenson also developed the concept of creating an 'apparent' light on dangerous reefs by indirect illumination and reflection from a parent lighthouse and installed a

HOLOPHOTAL SYSTEM.   81

densing the whole sphere of diverging rays into a single beam of parallel rays, without any unnecessary reflections or refractions. Part of the anterior hemisphere of rays (Figs. 52, 53) is intercepted and at once parallelised by the lens *L*,

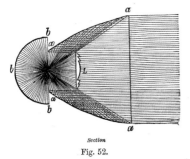

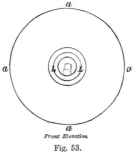

Section
Fig. 52.

Front Elevation
Fig. 53.

whose principal focus (*i.e.* for parallel rays) is in the centre of the flame, while the remainder is intercepted and made parallel by the paraboloid *a*, and thus the double agents in Fresnel's and Brewster's designs (**17, 19**) are dispensed with. The rays of the posterior hemisphere are reflected by the spherical mirror *b*, back again through the focus, whence passing onwards one portion of them falls on the lens and the rest on the paraboloid, so as finally to emerge in union with, and parallel to the front rays. This was the first instrument which intercepted and parallelised all the rays proceeding from a focal point by the minimum number of

Left: Catadioptric fixed holophote used at Peterhead Harbour in 1849 [Stevenson T. Lighthouse Construction and Illumination, 1881]

'beautiful and ingenious contrivance' ('Report … Stevenson's Paper on Dipping and Apparent Lights', 291) at Stornoway in 1851. Stevenson's crowning achievement was his 'azimuthal condensing system', which reduced the available light in some sectors of azimuth and optimized it in others. It was introduced at Isle Oronsay Lighthouse, Skye, (1857) to service Sleat Sound. He was assisted in some of the calculations required for his inventions by his cousin and lifelong friend Professor W. Swan and also by Professor P. G. Tait.

In 1848, Stevenson was elected a fellow of the Royal Society of Edinburgh, becoming its president in 1884. He was elected a member of the Institution of Civil Engineers in 1864, a fellow of the Geological Society in 1874, and was a founder member of the Scottish Meteorological Society in 1855, becoming its honorary secretary in 1871. Among the many and permanent contributions which he made to meteorology were the Stevenson screen for the protection of thermometers, designed in 1864 and now in universal use; the introduction in 1867 of the term 'barometric gradient'; and the means of ascertaining, by high and low level observations at Ben Nevis observatory and elsewhere, the vertical gradients for atmospheric pressure, temperature, and humidity.

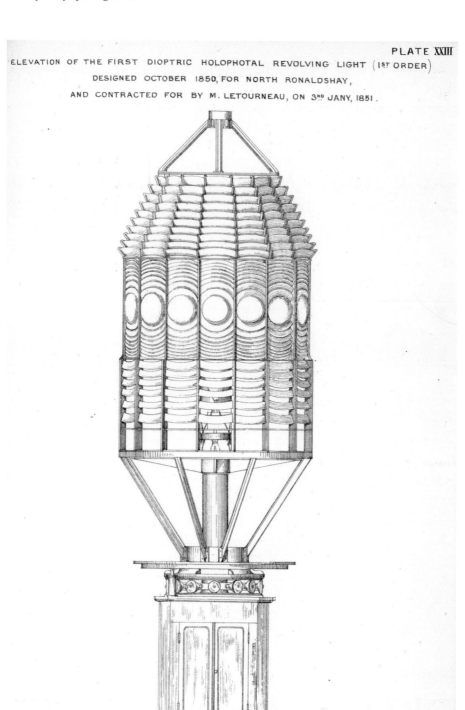

PLATE XXIII

ELEVATION OF THE FIRST DIOPTRIC HOLOPHOTAL REVOLVING LIGHT (1ST ORDER)
DESIGNED OCTOBER 1850, FOR NORTH RONALDSHAY,
AND CONTRACTED FOR BY M. LETOURNEAU, ON 3RD JANY, 1851.

Dioptric holophotal revolving light at North Ronaldsay, 1851
[Stevenson T. Lighthouse Construction and Illumination, *1881]*

On 8 August 1848, Stevenson married Margaret Isabella (1829–1897), daughter of the Revd Lewis Balfour, minister of Colinton. In her early and middle life Margaret suffered from chest problems. The couple had one child, the writer Robert Louis Stevenson. A devoted member of the Church of Scotland, Thomas Stevenson wrote several religious pamphlets including *Christianity Confirmed by Jewish and Heathen Testimony, and the Deductions from Physical Science* (1877; 2nd edn, 1879). He became ill with an enlarged liver in 1885; eventually he developed jaundice and died at his house, 17 Heriot Row, Edinburgh, on 8 May 1887. He was buried in the New Calton Cemetery, Edinburgh. He was survived by his wife who lived for a time in Samoa with her son.

Sources

W. Swan, 'Thomas Stevenson', *Proceedings of the Royal Society of Edinburgh*, 20 (1892–5), lxi–lxxviii · PICE, 91 (1887–8), 424–6 · R. L. Stevenson, *Records of a Family of Engineers* (1912) · R. L. Stevenson, *Memories and Portraits*, 8th edn (1898) · T. Stevenson, *Lighthouse Construction and Illumination* (1881) · T. Stevenson, *The Design and Construction of Harbours*, 3rd edn (1886) · C. Mair, *Star for Seamen: the Stevenson Family of Engineers* (1978) · J. M. Townson, 'Thomas Stevenson', *Shore and Beach*, 44/2 (1976), 3–12 · Business records of Robert Stevenson & Sons, NLS, Acc. 10706 · Private information (2004) · R. L. Stevenson, *Familiar Studies of Men and Books* (1886), v · 'Report of the Committee Appointed by the Royal Scottish Society of Arts on Thomas Stevenson's Paper on Dipping and Apparent Lights', *Transactions of the Royal Scottish Society of Arts*, 4 (1846–55), 291 · d. cert. · Family grave, New Calton Cemetery

Archives

NLS, corresp. with Scottish Society of Arts ·NLS, business records of Robert Stevenson & Sons, Acc. 10706

Likenesses

G. Reid, oils, Scot. NPG · Photograph (after G. Reid), Royal Society of Edinburgh · Photograph, repro. [in] Mair, Craig, *Star for Seamen: Stevenson Family of Engineers*

Wealth at death

£26,918 16s. 11d.: confirmation, 5 Aug 1887, CCI

The Stevenson family in Edinburgh in 1880. Robert Louis and Fanny (left),
Thomas and Margaret, and Lloyd Osborne (rear) (2, 81)

ROBERT LOUIS STEVENSON

1850–1894

The following is a reprint of my essay on Stevenson's three and a half years as 'a reluctant trainee civil engineer' in *Bright Lights: the Stevenson Engineers* (**2**, 99–109).

Although Louis never became a civil engineer, until he reached the age of 21 his parents hoped that he would follow in this family tradition and he was educated accordingly. In November 1867 he took a tentative step in this direction by enrolling as a student in the Arts faculty at Edinburgh University. It was not however until 1869–70 that he studied any engineering related subjects, namely mathematics and natural philosophy. He continued with these in 1870–71 together with the engineering classes of Professor Fleeming Jenkin (1833–85) which, being unable to follow, he refrained from attending.

By April 1871 after some three and a half years of dutifully, if increasingly half-heartedly, pursuing this career he felt unable to continue and announced his decision to give up engineering. This outcome was accepted with disappointment but also with 'wonderful resignation' [R. L. Stevenson. *Letters of* … (Eds: Bradford A. Booth and Ernest Mehew). Newhaven and London, 1995, 1, 209. Margaret Stevenson's diary, 8 April 1871] by his father who was no doubt recalling similar youthful tussles. They came to the understanding that Louis read for the bar, instead of writing literature, which his father considered no profession! Engineering's loss proved outstandingly to be literature's gain, but Louis's writings benefited immeasurably from his maritime engineering experience both in context and detail.

Some of Louis's published writings even related directly to engineering such as his paper on 'A New Form of Intermittent Light for Lighthouses', his *Memoir of Fleeming Jenkin, Records of a Family of Engineers* and an essay, 'The Education of an Engineer'. Louis and his father enjoyed corresponding on literary matters, each claiming to have improved some of the other's writings. Louis considered that he had 'materially helped to polish the diamond' of his father's presidential address to the Royal Society of Edinburgh in 1885 and 'ended by feeling quite proud of the paper as if it had been mine; the next time you have as good a one, I will overhaul it for the wages of feeling as clever as I did when I had managed to understand and helped to set it clear'. [R. L. Stevenson. *Letters of* … (Eds: Bradford A. Booth and Ernest Mehew). Newhaven and London, 1995, 5, 68].

Louis's intermittent light paper, read to the Royal Scottish Society of Arts on 27 March 1871, was a creditable effort and earned him a silver medal of the Society. It also prompted his jaunty farewell *To the Commissioners of Northern Lights*, which concluded with the thought that as a future advocate he might one day be a commissioner himself!

I send to you, commissioners,
A paper that may please ye, sirs,
(For troth they say it micht be worse
An' I believ't)
And on your business lay my curse
Before I leav't.

I thocht I'd serve wi' you, sirs, yince,
But I've thocht better of it since;
The maitter I will nowise mince,
But tell ye true:
I'll service wi' some ither prince,
An' no' wi' you.

I've no' been very deep, ye'll think,
Cam' delicately to the brink,
An' when the water gart me shrink,
Straucht took the rue,
An' didna stoop my fill to drink—
I own it true.

I kent on cape and isle, a light
Burnt fair an' clearly ilka night;
But at the service I took fright,
As sune's I saw,
An' being still a neophite
Gaed straucht awa'.

Anither course I now begin,
The weeg I'll cairry for my sin,
The court my voice sall echo in,
An'—wha can tell?—
Some ither day I may be yin
O' you mysel'.

[R. L. Stevenson. *Collected Poems.* (Ed: Janet Adam Smith), London, 1950, 101–102].

Louis's engineering education at Edinburgh University seems to have been characterised more by truancy and a very tolerant Professor Fleeming Jenkin than any serious acquisition of knowledge. Against his better judgement and after much pleading by Louis for a class attendance certificate first told him: 'It is quite useless for you to come to me, Mr. Stevenson. There may be doubtful cases, there is no doubt about yours. You have simply not attended my class' [R. L. Stevenson. *Memoir of Fleeming Jenkin*, 1912]. However, he later provided him with one containing a form of words for his father's eyes indicative of his having satisfactorily completed the class-work in engineering.

Louis wrote: 'I am still ashamed when I think of his shame in giving me that paper. He made no reproach in his speech, but his manner was the more eloquent; it told me plainly what dirty business we were on; and I went from his presence, with my certificate indeed in my possession, but with no answerable sense of triumph.' [R. L. Stevenson. *Memoir of Fleeming Jenkin*, 1912] There seems to have been no question of Louis graduating. This was the 'bitter beginning' of his great friendship with Jenkin of whom he wrote in 1885 'I never knew a better man' [Letter from Louis to Mrs Fleeming Jenkin, 12 June 1885. [In] R. L. Stevenson, *Works.* (Ed:

Sydney Colvin), London, Tusitala edition, 1923–24, XXXIII, 47] and on whom he bestowed to posterity a remarkable if not very comprehensive biography.

During the long summer vacations Louis gained practical experience of harbour and lighthouse engineering operations, particularly of pier construction at Anstruther and Wick in 1868. He was fascinated by the experience of sea diving but, in general, found the site work physically demanding and uncongenial. He wrote to his father from Anstruther on 2 July 1868: 'bring also my paint box. … I am going to try the travellers and Jennies, and have made a sketch of them and begun the drawing. After that I'll do the staging.' The 'travellers' were timber gantries that moved along the pier and ahead of its temporary end, on rails supported on piles at each side of the pier. The *Jennies* were cranes, which moved backwards and forwards transversely on top of the travellers. They were used for lifting and lowering stones into position [see p. 32 for the travellers and jennies at Wick].

> Tomorrow I will watch the masons at work at the pier foot and see how long they take to work that Fifeness stone you ask about: they get sixpence an hour; so that is the only datum required … It is awful how slowly I draw and how ill: I am not nearly done with the travellers and have not thought of the Jennies yet. When I'm drawing I find out something I have not measured, or, having measured, have not noted, or, having noted, cannot find; and so I have to trudge to the pier again, ere I can go further with my noble design.

Of his experience at Anstruther Louis wrote:

> What I gleaned, I am sure I do not know; but indeed I had already my own private determination to be an author; I loved the art of words and the appearances of life; and travellers and headers, and rubble, and polished ashlar [squared masonry], and pierres perdues [rubble stone], and even the thrilling question of string course [of masonry set out by string line], interested me only (if they interested me at all) as properties of some possible romance or as words to add to my vocabulary … though I haunted the breakwater by day, and even loved the place for the sake of the sunshine, the thrilling sea-side air, the wash of waves on the sea-face, the green glimmer of the divers helmets far below, and the musical chinking of the masons, my one genuine pre-occupation lay elsewhere, and my only industry was in the hours when I was not on duty.

Then northwards to Wick:

> Into the bay of Wick stretched the dark length of the unfinished breakwater, in its cage of open staging; the travellers (like frames of churches) over-plumbing all; and away at the extreme end, the divers toiling unseen on the foundation. On a platform of loose planks, the assistants turned their air mills; a stone might be swinging between wind and water; underneath the swell ran gayly; and from time to time a mailed dragon with a window glass snout came dripping up

the ladder … To go down in the dress, that was my absorbing fancy; and with the countenance of a certain handsome scamp of a diver, Bob Bain by name, I gratified the whim … Some twenty rounds below the platform, twilight fell. Looking up I saw a low green heaven mottled with vanishing bells of white; looking around, except for the weedy spokes and shaft of the ladder, nothing but a green gloaming.

Thirty rounds lower [at a depth of about 30 ft], I stepped off on the pierres perdues of the foundation; a dumb helmeted figure took me by the hand, and made a gesture (as I read it) of encouragement; and looking in at the creature's window, I beheld the face of Bain … how a man's weight, so far from being an encumbrance, is the very ground of his agility, was the chief lesson of my submarine experience … As I began to go forward with the hand of my estranged companion, a world of tumbled stones was visible, pillared with the weedy uprights of the staging: overhead, a flat roof of green: a little in front, the sea-wall, like an unfinished rampart.

And presently, in our upward progress, Bob motioned me to leap upon a stone … Now the block stood six feet high; it would have been quite a leap to me unencumbered; with the breast and back weights, and the twenty pounds upon each foot, and the staggering load of the helmet, the thing was out of reason. I laughed aloud in my tomb; and to prove to Bob how far he was astray, I gave a little impulse from my toes. Up I soared like a bird, my companion soaring at my side. As high as to the stone and then higher, I pursued my impotent and empty flight. Even when the strong arm of Bob had checked my shoulders, my heels continued their ascent; so that I blew out sideways like an autumn leaf, and must be hauled in hand over hand, as sailors haul in the slack of a sail, and propped upon my feet again like an intoxicated sparrow … Bain brought me back to the ladder and signed me to mount … Of a sudden, my ascending head passed into the trough of a swell. Out of the green I shot at once into a glory of rosy, almost sanguine light—the multitudinous seas incarnadined, the heaven above a vault of crimson. And then the glory faded into the hard, ugly daylight of a Caithness autumn, with a low sky, a gray sea, and a whistling wind.

Diving was one of the best things I got from my education as an engineer: of which however, as a way of life, I wish to speak with sympathy. It takes a man into the open air; it keeps him hanging about harbor-sides, which is the richest form of idling; it carries him to wide islands; it give him a taste of the genial dangers of the sea; it supplies him with dexterities to exercise; it makes demands upon his ingenuity; it will go far to cure him of any taste (if ever he had one) for the miserable life of cities. And when it has done so, it carries him back and shuts him in an office! From the roaring skerry and the wet thwart of the tossing boat, he passes to the stool and desk; and with a memory full of ships, and seas, and perilous headlands, and the shining pharos, he must apply his long-sighted eyes to the petty niceties of drawing, or measure his inaccurate mind with several pages of consecutive figures. He is a wise youth, to be sure, who can balance one part of genuine life against two parts of drudgery between four walls, and for the sake of the one, manfully accept the other. [R. L. Stevenson. 'The Education of an Engineer: More Random Memories.' *Scribner's Magazine*, November 1888, IV, 636–640].

In a letter to his mother on 20–21 September 1868, Louis wrote:

I was awakened by Mrs S. at the door [of the New Harbour Hotel, Pultneytown—now a Customs office]. There's a ship ashore at Shaltigoe! I got up, dressed and went out. The mizzled sky and rain blinded you … Some of the waves were 20 feet high. The spray rose 80 feet at the new pier … The thunder at the wall when it first struck—the rush along ever growing higher—the great jet of snow-white spray some 40 feet above you—and the 'noise of many waters', the roar, the hiss, the 'shrieking' amongst the shingle as it fell head over heels at your feet. I watched if it threw the big stones at the wall; but it never moved them.

[next day] The end of the [breakwater] work displays gaps, cairns of ten ton blocks, stones torn from their places and turned right round. The damage above water is comparatively little: what there may be below, 'ne sait pas encore'. The roadway is torn away, cross-heads broken, planks tossed here and there, planks gnawn and mumbled as if a starved bear had been trying to eat them, planks with spales lifted from them as if they had been dressed with a ragged plane, one pile swaying to and fro clear of the bottom, the rails in one place sunk a foot at least. This was not a great storm, the waves were light and short. Yet when we are [were] standing at the office, I felt the ground beneath me quail as a huge roller thundered on the work at the last year's cross-wall …

[To] appreciate a storm at Wick requires a little of the artistic temperament which Mr.T.S.C.E. [Thomas Stevenson, Civil Engineer] possesses … I can't look at it practically however: that will come I suppose like gray hair or coffin nails … Our pole is snapped: a fortnight's work and the loss of the Norge schooner all for nothing!—except experience and dirty clothes.

[R. L. Stevenson. *Letters of …* (Eds: Bradford A. Booth and Ernest Mehew), Newhaven and London, 1995, I, 156–158].

[Stevenson acted as a trainee engineer at Wick for 40 days from 28 August 1868 when the breakwater had reached about 70 per cent of its planned length or about 320m. from the shore. This was the maximum length actually achieved as from October 1868 heavy seas destroyed lengths of breakwater on at least four occasions leading to its shortening and eventual abandonment in 1877. He dubbed the breakwater, 'the chief disaster of my father's life … the sea proved too strong for man's arts: and after expedients hitherto unthought of and on a scale hyper-cyclopean, the work must be destroyed, and now stands a ruin in that bleak, Godforsaken bay' (5)(13)].

Wick Breakwater c. 1865. Note the 'travellers and jennies'
[the travelling cranes above the timber staging] being supplied with
stone by railway from the local South Head Quarry. © The Wick Society (2, 101)

In a letter to his cousin Bob Stevenson, Louis paints an indelible picture of the mail coach journey from Wick, by night, to the most northerly railway terminus then at Golspie (about 50 miles to the south). [The railway did not reach Wick until 1874].

Mail coach at Wick in 1874 as familiar to Louis. © The Wick Society (2, 103)

The Wick Mail then, my dear fellow, is the last Mail Coach within Great Britain, whence there comes a romantic interest that few could understand. To me, on whose imagination positively nothing took so strong a hold as the Dick Turpins and Claude Duvals of last century, a Mail was an object of religious awe. I pictured the long dark highways, the guard's blunderbuss, the passengers with three-cornered hats above a mummery of great-coat and cravat; and the sudden 'Stand and deliver!'—the stop, the glimmer of the coach lamp upon the horseman—Ah! we shall never get back to Wick.

All round that northern capital of stink and storm there stretches a succession of flat and dreary moors absolutely treeless, with the exception of above a hundred bour-trees [elders] beside Wick, and a stunted plantation at Stirkoke, for the distance of nearly twenty miles south. When we left to cross this tract, it was cloudy and dark. A very cold and pertinacious wind blew with unchecked violence, across these moorlands. I was sick sleepy [?], and drawing my cloak over my face set myself to doze. Mine was the box-seat, desirable for the apron and the company of the coachman, a person in this instance enveloped in that holy and tender interest that hangs about the 'Last of the Mohicans' or the 'Derniers Bretons'.

And as this example of the loquacious genus coachman was more than ordinarily loquacious I put down my hood again and talked with him. He had a philosophy of his own, I found, and a philosophy eminently suited to the needs of his position. The most fundamental and original doctrine of this, was as to what constitutes a gentleman. It was in speaking of Lockyer of Wenbury that I found it out. This man is an audacious quack and charlatan, destined for aught, I know, to be the Cagliostro [Italian charlatan] of the British Revolution; and, as such, Mr Lockyer is no favourite of mine: I hate quacks, not personally (for they are not men of imagination like ourselves?) but because of their influence; so I was rather struck on hearing the following. 'Well sir', said the coachman, 'Mr Lockyer has always shown himself a perfect gentleman to me, sir—his hand as open as you'll see, sir!' In other words, half-a-crown to the coachman!

As the pleasures of such philosophical talk rather diminished and the slumber increased, I buried my face again. The coach swayed to and fro. The wind battled and roared about us. I observed the difference in sounds—the rhythmic and regular beat of the hoofs as the horses cantered up some incline, and the ringing, merry, irregular clatter as they slung forward, at a merry trot, along the level.

First stage: Lybster. A Roman catholic priest travelling within, knowing that I was delicate, made me take his seat inside for the next stage. I dozed. When I woke, the moon was shining brightly. We were off the moors and up among the high grounds near the Ord of Caithness. I remember seeing a curious thing: the moon shone on the ocean, and on a river swollen to a great pool and between stretched a great black mass of rock: I wondered dimly how the river got out and then to doze again. When next I wake, we have passed the low church of Berridale, standing sentinel on the heathery plateau northward of the valley, and are descending the steep road past the Manse: I think it was about one: the moon was frosty but gloriously clear.

In another minute—

Second stage: Berridale. And of all lovely places, one of the loveliest. Two rivers run from the inner hills, at the bottom of two deep, Killiecrankie-like gorges, to meet in a narrow bare valley close to the grey North Ocean. The high Peninsula between and the banks, on either hand until they meet, are thickly wooded—birch and fir. On one side is the bleak plateau with the lonesome little church, on the other the bleaker, wilder mountain of the Ord. When I and the priest had lit our pipes, we crossed the streams, now speckled with the moonlight that filtered through the trees, and walked to the top of the Ord. There the coach overtook us and away we went for a stage, over great, bleak mountains, with here and there a hanging wood of silver birches and here and there a long look of the moonlit sea, the white ribbon of the road marked far in front by the newly erected telegraph posts. We were all broad awake with our walk, and made very merry outside, proffering 'fills' of tobacco and pinches of snuff and dipping surreptitiously into aristocratic flasks and plebeian pint bottles.

Third stage: Helmsdale. Round a great promontory with the gleaming sea far away in front, and rattling through some sleeping streets that shone strangely white in the moonlight, and then we pull up beside the Helmsdale posting-house, with a great mountain valley behind. Here I went in to get a glass of whiskey and water. A very broad, dark commercial said: 'Ha! do you remember me? Anstruther?' I had met him five years before in the Anstruther commercial room, when my father was conversing with an infidel and put me out of the room to be away from contamination; whereupon I listened outside and heard the man say he had not sinned for seven years, and declare that he was better than his maker. I did not remember him; nor did he my face, only my voice. He insisted on standing me the whiskey "for auld lang syne"; and he being a bagman, it was useless to refuse.

Then away again. The coachman very communicative at this stage, telling us about the winter before, when the mails had to be carried through on horseback and how they left one of their number sticking in the snow, bag and all I suppose. The country here was softer; low, wooded hills running along beside the shore and all inexpressibly delightful to me after my six months of Wick barrenness and storm.

Fourth stage: name unknown. O sweet little spot, how often I have longed to be back to you! A lone farm-house on the sea-shore, shut in on three sides by the same, low, wooded hills. Men were waiting for us by the roadside, with the horses—sleepy, yawning men. What a peaceful place it was! Everything steeped in the moonlight, and the gentle plash of the waves coming to us from the beach. On again. Through Brora, where we stopped at the Post-Office and exchanged letter-bags through a practicable window-pane, as they say in stage directions. Then on again. Near Golspie now, and breakfast, and the roaring railway. Passed Dunrobin, the dew-steeped, tree-dotted park, the princely cluster of its towers, rising from bosky plantations and standing out against the moon-shimmering sea—all this sylvan and idyllic beauty so sweet and new to me! Then the Golspie Inn, and breakfast and another pipe, as the morning dawned, standing in the verandah. And then round to the station to fall asleep in the train …

[R. L. Stevenson. *Letters of …* (Eds: Bradford A. Booth and Ernest Mehew), Newhaven and London, 1995, I, 169–172].

In June 1869 Louis accompanied his father on a voyage of inspection of lighthouses in Orkney and Shetland in the lighthouse steamer and provided his mother with a detailed account of his 'sore journeying and perilous peregrination'. For example:

> … we sighted North Unst Lighthouse, the most northerly dwelling house in Her Majesty's dominion. The mainland rises higher, with great seams and landslips; and from the norwestern corner runs out a string of shelving ledges, with a streak of green and purple seaweed and a boil of white foam about their feet. The lighthouse stands on the highest—190 feet above the sea; … the reefs looked somewhat thus … [see sketch]

Louis's sketch of Muckle Flugga in 1869 [R. L. Stevenson. Letters of …
*(Eds: Bradford A. Booth and Ernest Mehew), 1995. **I**. 181]. (2, 85)*

We were pulled into the creek shown in the picture between the lighthouse and the other rock, down the centre of which runs a line of reef … This is very narrow, little broader than a knife edge; but its ridge has been cut into stone steps and laid with iron grating and railed with an iron railing. It was here that we landed, making a leap between the swells at a rusted ladder laid slant-wise against the raking side [of] the ridge. Before us a flight of stone steps led up the 200 feet to the lighthouse in its high yard-walls across whose foot the sea had cast a boulder weighing 20 tons. On one side is a slippery face of clear sound rock and on the other a chaos of pendulous boulder and rotten stone. On either side there was no vegetation save tufts of sea-pink in the crevices and a little white lichen on the lee faces.

[R. L. Stevenson. *Letters of* ... (Eds: Bradford A. Booth and Ernest Mehew), Newhaven and London, 1995, 8, 181].

During his summer vacation of 1870 Louis spent three weeks on the isle of Erraid [off Mull], from which he visited Dhu Heartach Lighthouse then under construction 15 miles to the south-west. Two years later, although then pursuing his law classes at University, he managed in the summer to produce a lively account of the project:

> Even before the work was sanctioned, he wrote, Dhu Heartach had given the engineers a taste of its difficulties. Although the weather was fine, Messrs Stevenson failed to effect a landing and had to send in their preliminary report based merely upon what they could see from the deck of the vessel. But even this had not prepared them for the continual difficulty and danger which accompanied every landing from the beginning to the end of the work. Favoured by the smooth egg-shaped outline of the rock, which is about [130] feet broad, [240] feet long and 35 feet above high water at its summit, the swell breaks at the one end, runs cumulating round either side, and meets and breaks again at the opposite end, so that the whole rock is girdled with broken water. There is no sheltered bight. If there be anything to aggravate the swell, and it is wonderful what a little thing it takes to excite these giant-waters, landing becomes impossible ... The probability is that the very height of Dhu Heartach rock, by causing the waves to rise, is what makes them so dangerous at a considerable elevation; in short the destructive character of a wave as regards level depends upon the relation between the height of the wave, the height and contour of the obstacle and the depth of water in which it acts.
>
> The first object of the Messrs Stevenson was to erect a temporary barrack for the residence of the workmen ... it was decided that the structure—a framework 35 feet high, supporting a plated cylinder or drum 20 feet high and divided into two stories—should consist entirely of malleable iron [see figure] ... the shore station was placed on Isle Earraid. On the twenty-fifth of June ... they first took possession of the rock and disturbed the seals, who had been its former undisputed tenants. ...
>
> The work during this first season was much interrupted. Even when a landing was effected, the sea rose so suddenly and there was such a want of appliances upon the Rock itself, that the men had sometimes hard [had?] enough ado to get off again. ... The season which began so late closed finally on the third of September; and the first tier of the barrack framework was left unfinished.
>
> All the winter of 1867–68, a band of resident workmen were carrying on the shore station with its bothies, cottages, quarry and the workyard where every stone was to be cut, dressed, fitted and numbered before being sent out to

the rock to be finally built into the tower; and on the fourteenth of April, the Dhu Heartach *steamer* came back to her moorings in Earraid Sound. The result proved that she was too early; for there was no landing at all in April; only two in May, giving between them a grand total of two hours and a half upon the rock; and only two once again in June. In July there were 13, in August ten, in September 11: in all, 38 landings in five months. ...

On the twentieth of August, the malleable iron barrack was so far advanced and the weather gave such promise of continuing fine, that Mr Alan Brebner C.E. (of Edinburgh) and 13 workmen landed on the rock and took up their abode in the drum ... A sudden gale however sprung up and they could not be communicated with till the 26th [August 1868] during the greater part of which time the sea broke so heavily over the rock as to prevent all work and during the height of the storm the spray rose far above the barrack and the sea struck very heavily on the flooring of the lower compartment which is 35 feet above the rock and 56 feet above high water mark.

The third season, that of 1869, saw the work properly commenced. The barrack had come scatheless through the winter, and the master-builder Mr Goodwillie with between 20 and 30 workmen took up his abode there, on the twenty-sixth of April. On Isle Earraid, there was a good quarry of granite, two rows of sheds, two travelling cranes, railways to carry the stones, a stage on which, course after course, the lighthouse was put experimentally together and then taken down again to be sent piece-meal out to the rock, a pier for the lighters [stone carrying boats], and a look out place furnished with a powerful telescope by which it could be observed whether the weather was clear [and] how high the sea was running on Dhu Heartach and so judge whether it were worthwhile to steam out on the chance of landing. In a word, there was a stirring village of some [50] souls, on this island which, four years before, had been tenanted by one fisherman's family and a herd of sheep.

The life in this little community was highly characteristic. On Sundays only, the continual clink of tools from quarry and workyard came to an end, perfect quiet then reigned throughout the settlement, and you saw workmen leisurely smoking their pipes about the green enclosure, and they and their wives wearing their Sunday clothes (from association of ideas, I fancy) just as if they were going to take their accustomed seats in the crowded church at home. As for the services at Earraid, they were held in one of the wooden bothies, the audience perched about the double tier of box beds or gathered round the table. Mr Brebner [The Engineer] read a sermon and the eloquent prayer which was written specially for the Scottish Lighthouse service, and a voluntary band and precentor led the psalms. Occasionally, a regular minister came to the station, and then worship was held in the joiner's shop. ...

In fine weather, before the sun had risen behind Ben-More, the Dhu Heartach steamed out of the bay towing a couple of heavy, strong-built lighters laden with the dressed and numbered stones. It was no easy or pleasant duty to be steersman in these lighters, for what with the deck-cargo and the long heavy swell, they rolled so violently that few sailors were able to stand it. Dhu Heartach itself on some such calm, warm summer day presented a strange spectacle. This small black rock, almost out of sight of land in the fretful, easily-irritated sea, was a centre of indefatigable energy.

The whole small space was occupied by men coming and going between the lighters or the barrack and the slowly-lengthening tower. A steam winch and inclined plane raised the stones from the water's edge to the foot of the building; and it was a matter of no little address and nicety, to whip one of these great two-ton blocks out of the lighter, [see sketch] as it knocked about and rolled gunwale-under in the swell, and bring it safely up to the tower, without breaking it or chipping off some corner that would spoil the joint. Then, there would come the dinner horn; and the noise was incontinently quieted, there was no more puffing of the steam engine or clink of the mallets on the building; the men sat scattered in groups over their junk [salt meat] and potatoes and beer. ...

By the end of this season, the tower had reached the height of eight feet, four inches... But the heaviest end of the work was now over. In the fourth season, 1870, there were 62 landing days, and the white tower soon began to top its older brother the iron barrack. By the end of that season, it was 48 feet high, the last stone was laid next summer, and, during the present summer, the lantern and internal fittings have also been brought to completion. Before the end of 1872, the light will have been exhibited. For the tower ... Messrs Stevenson adopted:

> the form of a parabolic frustum, to a hundred and seven and a half feet high, 36 feet in diameter at the base and 16 feet at the top all built of granite ... The entire weight of masonry is 3,115 tons. ... The light will be fixed dioptric with a range of 18 miles in addition to which there is machinery which rings during fogs a hundredweight bell.

Anticipating possible future discomforts arising from the difficulty of landing supplies, he concludes:

> Shortly before the light was first exhibited, [at nearby Skerryvore Lighthouse] ... a long track of storms extending over seven weeks prevented the tender from getting near the reef; and before the weather had moderated, the stock of tobacco in the tower was quite exhausted. On the morning of this catastrophe, the [workmen] ceremoniously broke their pipes and put up a chalk inscription over the mantle shelf in the kitchen: 'Such-and-such a date, Tobacco done—Pipes Broken.' Let us hope that no such 'memor querela' may ever be read over the chimney of Dhu Heartach.

[R. L. Stevenson. *The New Lighthouse on the Dhu Heartach Rock, Argyllshire* (Ed: Roger G. Swearingen), St Helena, California, 1995, 1–21].

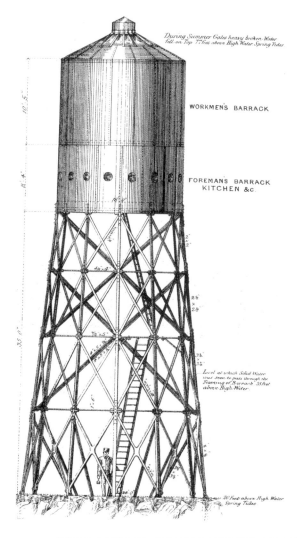

WORKMENS BARRACK ON THE ROCK.

During Summer Gales heavy broken Water fell on Top 7 Feet above High Water Spring Tides

WORKMEN'S BARRACK

FOREMANS BARRACK KITCHEN &c.

Level at which Solid Water was seen to pass through the Framing of Barrack 35 feet above High Water

80 feet above High Water Spring Tides

*Left: Workmen's barrack Dhu Heartach Rock [T. Stevenson] (**2**, 106)*

*Below: Sketch of lifting stones from a lighter by derrick to a truck at the bottom of the inclined plane at Dhu Heartach in 1870 (**2**, 117, by David Alan (Louis's 16 year old cousin)*

It is doubtful whether Louis's delicate health would have allowed him to become a successful engineer even if he had had the inclination and had persevered with his engineering education. Fortunately for posterity he gave reign to what was undoubtedly his greater talent, but it was not to prove a decision which was to bring him complete peace of mind. At the age of 43, in the last year of his life, he wrote to W. H. Low in a fit of depression, that his literary achievements had been inadequate for:

> the top flower of a man's life ... Small is the word; it is a small age and I am of it. I could have wished to be otherwise busy in this world. I ought to have been able to build lighthouses and write 'David Balfours' too. Hinc illae lacrymae [hence these tears]. [R. L. Stevenson. *Letters of* ... (Eds: Bradford A. Booth and Ernest Mehew). Newhaven and London, 1995, 8, 235]

Far left: David Alan Stevenson. *(2, 151)*

Left: [David] Alan Stevenson. [Family source]

Below: Charles Alexander Stevenson at D. and C. Stevenson's office, 84 George Street, Edinburgh, c. 1920. (2, 170)

DAVID ALAN STEVENSON

1854–1938

CHARLES ALEXANDER STEVENSON

1855–1950

(DAVID) ALAN STEVENSON

1891–1971

David Alan Stevenson (1854–1938), civil engineer, and his business partner, Charles Alexander Stevenson (1855–1950), were sons of David Stevenson and born in Edinburgh (at 8 Forth Street and 20 Royal Terrace respectively) on 21 July 1854 and 23 December 1855. They were both educated at Scott's Preparatory School, Edinburgh Academy, and Edinburgh University. David (usually known as David A. Stevenson) graduated BSc in 1875 and Charles likewise in 1877. Robert Louis Stevenson (1850–1894) was their cousin. Charles's son, (David) Alan Stevenson (1891–1971), civil engineer, usually known as D. Alan Stevenson, was born in Edinburgh at 9 Manor Place on 7 February 1891. Educated at Edinburgh Academy and Edinburgh University, he spent most of his professional life working with his father and uncle.

Early work

David and Charles Stevenson began their rigorous engineering training in the offices of D. and T. Stevenson on the second floor of 84 George Street, Edinburgh, during school vacations in 1868 and 1869, and entered into their formal three-year apprenticeships in 1875 and 1877. From 1875 to 1880 the firm was engaged on lighthouses at the Isle of Man and Holy Island, Arran, and on harbour work at Port Seton, Boddam, St Monance, Burnmouth, Gourock, Findochtie, Anstruther, and Broadford in Skye. Other work included preparations for deepening the Clyde from Port Glasgow seawards to 18 feet at low water for the Clyde Lighthouses Trust, which, in addition to lighthouse responsibilities, had a navigational remit for this length of river. David was an able technical writer and read his first paper 'Dhu Heartach Lighthouse' to a meeting at the Institution of Civil Engineers in 1876. This paper, based on the firm's reports and his teenage visits to the isolated rock during the lighthouse's construction, was awarded an Institution Manby premium prize. In 1878 David and the firm's most able senior assistant for many years, Alan Brebner (1826–1890), were taken into the partnership; Charles followed in 1886.

During the 1880s the firm undertook sewerage, harbour works, inspection of water-supply schemes for the Board of Supervision, and maintenance of North Esk Reservoir in the Pentland Hills. It was involved in the preparation of a major Forth and Clyde ship canal scheme that was level between sea-locks via Loch Lomond and Loch Long, which was never implemented, and it opposed the proposed Forth Bridge, to ensure an adequate navigational height, and the original Manchester ship canal scheme. For the Northern Lighthouse Board, the Stevensons introduced Courtenay whistling and Pintsch gaslight buoys. The latter were developed by Charles in 1880–81 and were integral to important experiments on the relative merits of oil, gas, and electricity as illuminants.

Other work for the board included the design and construction of the Ailsa Craig Lighthouse and foghorns, new lights in Orkney and Shetland, the North Carr lightship, and the introduction of electric light at the Isle of May Lighthouse. In 1884, with his brother David's retirement through ill health, Thomas Stevenson had become the board's sole engineer, but when his health began to fail in 1885 his nephew the younger David was appointed to act jointly with him. He was sole Engineer from Thomas's death in 1887 until 1938.

D. AND C. STEVENSON

1890–1936

In 1890 the firm became D. and C. Stevenson and until the First World War it was involved with navigational improvements of the Lune, Forth, and Manchester ship canals, the Forth and Clyde canal scheme, Oban sewerage, the branch railway from Longniddry to Gullane, deepening the lower Clyde to 27 feet, and numerous harbours in Scotland. Work for the Northern Lighthouse Board included lighthouse stations at Fair Isle, Rattray Head, Sule Skerry, Stroma, Scarinish, Noup Head, Flannan Island, Tiumpan Head, Bass Rock, Barns Ness, Killantringan, Hyskeir, Elie-Ness, Firths Voe, Neist Point, Rubha Reidh, Crammag Head and Maughold Head on the Isle of Man, and several lightships. Charles designed and implemented in 1910 at Platte Fougère, Guernsey, a remotely controlled and electrically operated acetylene light and foghorn installation—a precursor of the later lighthouse control system.

Other work included the installation of 'group flashing lights' at the Mull of Kintyre Lighthouse and the provision of many minor lights. The introduction of Sir James Chance's incandescent burners to Scottish lighthouses in 1903 and the wireless activation of equipment in 1914 were innovations of considerable significance by the Stevensons that became widely adopted in the lighthouse service. Except when specifically requested, David restricted his horizon to the Northern Lighthouse Service while Charles dealt with the firm's other work, but they collaborated on reports. This combination of talents, with Charles concentrating on equipment innovation and river engineering and his brother on the Board's work, worked well, and ensured the success of the firm until its dissolution in 1936. The brothers acted jointly as Engineers to the Clyde Lighthouses Trust and the Fishery Board of Scotland. In 1875 the most powerful light on the Scottish coast was of 44,500 candlepower. By 1901 there were several lights of over 100,000 candlepower, and the Isle of May electric light was of 3,000,000 candlepower. This increase in power was achieved by means of long focal distance apparatus that the Stevensons designed together, and the introduction of Charles's equiangular prisms which condensed the light into a narrower and more brilliant beam than previously. Before the turn of the nineteenth-century, David and Charles Stevenson had become leading authorities in their field, and during their partnership they acted as consulting engineers to several colonial and foreign lighthouse authorities.

During the same period Charles Stevenson pioneered the development of wireless communication and corresponded with Oliver Heaviside, Sydney Evershed, and other experimenters. In 1893–94 he conducted experiments in telephonic communication by electrical induction using coils, in which he successfully received speech over a distance of about half a mile at Murrayfield, Edinburgh. These experiments were carried out with a view to establishing a communication between Muckle Flugga Lighthouse and the Shetland mainland, but the need for this was questionable and it was never implemented. Charles is said to have anticipated Marconi, Heaviside, Preece and others in wireless transmission of speech. The apparatus did, however, have the drawback of being inconvenient to use because of the large diameter of the coil layout required (200 yards to transmit over half a mile) and its limited range. Charles Stevenson also invented the 'leader cable' for guiding vessels by means of an electric submarine cable laid on the ocean bed, a system eventually developed and installed at several large ports in Europe and the USA. It was used during the First World War in the North Sea off Harwich to guide vessels through minefields, and both these inventions formed the subject of papers read to the Royal Society of Edinburgh in 1894 and 1893. He also invented the automatic acetylene fog-gun signal.

Among Charles Stevenson's later innovations, jointly with his son, was the 'talking beacon' for use by ships in fog, which enabled the position of a ship to be plotted on board from synchronized radio and sound signals through the air. The radio signal provided the direction of the beacon and the distance from it was obtained by measuring the sound travel time from hearing a fog-signal detonation at the beacon on the ship's radio to its time of arrival at the ship through the air. The equivalent distance was stated on the radio in miles and cables at every cable (one-tenth of a nautical mile). This precursor of modern radar and satellite navigation was developed for the Clyde Lighthouses Trust and installed at the Cumbrae and Cloch lighthouses in 1929 and 1939. In 1931 this invention earned father and son the Thomas Gray award of the Royal Society of Arts.

In 1914, after having graduated with a BSc degree at Edinburgh University and serving his apprenticeship under his father, D. Alan Stevenson became an assistant in the firm. In 1919 he was taken into partnership. By the 1920s the period of great development of harbours, lighthouses, and navigational river deepenings in Britain had passed, and the nation was in the throes of economic recession. Nevertheless, the firm still kept busy in a smaller way on such works as further deepening the lower Clyde, sewerage and water supply projects and, more particularly, on the maintenance of lighthouse stations and modernization of equipment.

In 1926–27, for the government of India, Alan Stevenson inspected more than 100 lighthouses and advised on the organization of a centralized lighthouse service. In the 1930s, then the most physically active member of the firm, he skilfully superintended the deepening of the Clyde from Port Glasgow westwards, which enabled the *Queen Mary* to go to sea in 1936. This year turned out to be a fateful one for the family firm. According to Charles's granddaughter Jean Leslie, Alan Stevenson had grown impatient with the uncertainty of obtaining the post of engineer to the Northern Lighthouse Board still held by David Stevenson, then aged 81. This led to a difference that resulted in David Stevenson's withdrawal from D. and C. Stevenson, thus ending the firm's long-standing association with the board.

The D. and C. Stevenson firm was reformed in 1936, as A. and C. Stevenson, at 90a George Street, Edinburgh. As the NLB work continued with David Stevenson who was not now a partner, the new firm's workload was much reduced, relying principally on the family commission as joint Engineers to the Clyde Lighthouses Trust. Four years later Charles Stevenson retired, after which Alan Stevenson continued to act solely as the Trust's Engineer. His retirement in 1952 marked the end of nearly a century and a half of continuous service by Stevenson engineers to the Trust. In 1945 he read what is now a historically valuable paper on the engineering work of the Trust to the Institution of Engineers and Shipbuilders in Scotland.

Learned societies, personal life, outside activities

In 1886 and 1891 respectively, David and Charles Stevenson became members of the Institution of Civil Engineers, at which in 1887 David Stevenson read a paper on the Ailsa Craig Lighthouse station, winning him a Telford premium prize. He also addressed the Institution of Mechanical Engineers on the installation of electric light at the Isle of May. The following year he and Charles Stevenson revised, updated, and published the third and most comprehensive edition of their father's influential textbook—*Canal and River Engineering*. From 1894 to 1897 David Stevenson was an extramural examiner at the University of Edinburgh. He also gave evidence before royal commissions, parliamentary committees, and in important legal cases.

Charles Stevenson became a regular correspondent to *Nature* and other scientific periodicals on such diverse matters as the 1880 earthquake in Scotland, seismography, river discharge formulae, wind velocity and dioptric lenses. In 1887 and 1888 he read papers on dredging the Clyde and on a dipping or fog apparatus for electric light in lighthouses to the Institution of Mechanical Engineers and the Institution of Civil Engineers. To the latter in 1894, he read a paper on his important innovation of equiangular prism refractors for brightening dioptric lights. David and Charles Stevenson were elected fellows of the Royal Society of Edinburgh in 1884 and 1886 respectively. They were also members of the Royal Scottish Society of Arts and the Highland and Agricultural Society. In 1893 Charles Stevenson became an associate member of the Institution of Electrical Engineers. In 1919 Alan Stevenson continued in the family tradition by being elected a fellow of the Royal Society of Edinburgh, and in 1925 a member of the Institution of Civil Engineers.

Charles Stevenson married, on 19 January 1889, Margaret (1863–1945), daughter of Lieutenant-General John P. Sherriff. On 21 January 1892 David Stevenson married Dorothy Anne Roberts (*c.*1862–1945). The brothers' outside interests included golf, archery with the Royal Company of Archers, and skating with the Edinburgh Skating Club (for whom Charles Stevenson wrote the article 'Statics and Dynamics of Skating' in *Nature* [20 January 1881], which is also believed to have been separately printed as a pamphlet). The second edition of a booklet entitled *Skating Diagrams: Drawn by Charles A. Stevenson, C.E.* was published in 1881 at the expense of the Glasgow Skating Club. Later, Alan Stevenson also took up these interests and became the last secretary of the Edinburgh Skating Club. He was also a fellow of the Royal Scottish Geographical Society and served as its honorary treasurer. He married Jessie Laura Margaret MacLellan (1897–1975) on 5 June 1923.

David Stevenson served as Engineer to the Northern Lighthouse Board until his retirement on 31 March 1938, which ended 151 years of family service. He died on 11 April 1938 at his home, Troqueer, Kingsknowe, Edinburgh, and was buried in the Dean cemetery, Edinburgh. Charles Stevenson, the most inventive member of the family, with the possible exception of Thomas Stevenson, was described by his Royal Society of Edinburgh biographer as: 'a man of great intellectual acumen, but for work of this sort [engineering] he had a very special natural aptitude—a sort of instinctive grasp of how nature would work in the waves and winds and tides. His personal character had a great charm. Kindliness, gentleness and tolerance are characteristics that come to one's mind in recalling him, and permeating them all there was a natural unaffected simplicity and absence of sophistication.' (Johnstone)

He died on 9 May 1950 at his home, 29 Douglas Crescent, Edinburgh, and was also buried in the Dean Cemetery.

By the time of his retirement in 1952 Alan Stevenson had become increasingly involved in research for various historical pursuits, an interest which can be traced back to his article 'Early Scottish Lighthouses' in *Chambers Journal* (1917). He published his first full-length book in 1949—an account of Robert Stevenson's *English Lighthouse Tours, 1801, 1813, 1818*. This was followed by his definitive work, *The Triangular Stamps of Cape of Good Hope* (1950), for which he was awarded the Crawford medal of the Royal Philatelic Society and, his authoritative *The World's Lighthouses Before 1820* (1959).

Alan Stevenson is now remembered not so much for his engineering achievements but as a diligent custodian of the family's business records from about 1800, now at the National Library of Scotland (MS Acc. 10706), and as the nation's, if not the world's, foremost lighthouse historian of his day. His detailed unpublished notes and articles relating to the life and work of his engineering forebears proved an invaluable source of information to his biographers (Mair, Leslie, and Paxton). It was said of Alan Stevenson by the general manager of the Northern Lighthouse Board that: 'His was a life dedicated to lighthouses; his interest in them never flagged and a visitor to his bedside during his last illness will always remember his keen desire to discuss the latest developments. We in the Lighthouse Services of the world have lost not only an acknowledged authority but a true and staunch friend.' (Robertson)

He died on 22 December 1971 at 25 Belgrave Crescent, Edinburgh, and was buried in the Dean Cemetery.

Conclusion

With Alan Stevenson's death the family dynasty of civil engineers came to an end. It had begun with the appointment of Thomas Smith as Engineer to the Northern Lighthouse Board in 1787 and spanned five generations and 165 years of professional practice.

The family's main contributions to society were the improvement of maritime safety, the facilitation of trade by means of the design and erection of more than 200 lighthouses, improving the illumination of many of the world's lighthouses, and navigational or drainage improvements to numerous British harbours and most major rivers from the Ouse northwards to the Dornoch Firth. The firm's private work played a major part, as can be gauged from the fact that at the height of its success in the third quarter of the nineteenth-century only about one-quarter of its profits were attributed to Northern Lighthouse Board work.

Seven of the Stevenson engineers had the remarkable distinction of being members of both the Institution of Civil Engineers (1828–1971) and fellows of the Royal Society of Edinburgh (1815–1971). Their innovative work was often at the frontiers of technology and practice, and collectively they produced more than 200 professional publications including several influential textbooks. Overall, it was an outstanding achievement for a small firm, and one that has earned the Stevensons a place among the nation's most notable civil engineering families. Robert Louis

Stevenson's eloquent family reminiscence of their beacons and towers around the coast forms a fitting finale:

In the afternoon of time
A strenuous family dusted from its hands
The sand of granite, and beholding far
Along the sounding coast its pyramids
And tall memorials catch the dying sun,
Smiled well content,

[*Poems*, 'Underwoods', XXXVIII, *c.*1885]

Sources

J. Leslie and R. Paxton, Bright Lights: the Stevenson Engineers, 1751–1971 (1999) · C. Mair, Star for Seamen: the Stevenson Family of Engineers (1978) · J. D. Gardner, Proceedings of the Royal Society of Edinburgh, 58 (1937–8), 280–82 · R. W. Johnstone, 'Charles Alexander Stevenson, BSc, MICE', Year Book of the Royal Society of Edinburgh (1950–51) · A. Robertson, International Lighthouse Authority Bulletin (April 1972) · C. Mair, David Angus: the Life and Adventures of a Victorian Railway Engineer (1989) · Private information (2004) [family] · d. cert. [David Alan Stevenson] · d. cert. [Charles Alexander Stevenson] · d. cert. [(David) Alan Stevenson]

Archives

NAS, Northern Lighthouse Board records · National Monument Record of Scotland, RCAHMS. · NLS, business records of Robert Stevenson and sons, civil engineers, MS Acc. 10706

Likenesses

Photograph, c.1920 (Charles Alexander Stevenson), repro. [in] Leslie and Paxton, Bright Lights: the Stevenson Engineers · Photograph, c.1932 ((David) Alan Stevenson), repro. [in] Leslie and Paxton, Bright Lights: the Stevenson Engineers · Photograph, repro. [in] Mair, Star for Seamen: the Stevenson Family of Engineers · Photograph, repro. [in] Leslie and Paxton, Bright Lights: the Stevenson Engineers · Photograph (Charles Alexander Stevenson), repro. [in] Mair, Craig, Star for Seamen: the Stevenson Family of Engineers · Photograph ((David) Alan Stevenson), repro. [in] Mair, Craig, Star for Seamen: the Stevenson Family of Engineers

Wealth at death

£160,444—(David) Alan Stevenson: confirmation, 1972, NAS, SC 70/1/2029/54–60.

CHRONOLOGICAL LIST OF STEVENSON PRIVATE FIRMS

Robert Stevenson	March? 1811 – c. 1832
Robert Stevenson & Son	c. 1832 – May 1838
Robert Stevenson & Sons	May 1838 – 2 August 1849
D. & T. Stevenson	1849 – 1890*
D. & C. Stevenson	1890 – October 1936
A. & C. Stevenson	October 1936 – 1940
D. Alan Stevenson	1940 – 1952

* The firm continued to practise under this name after the deaths of David and Thomas. Alan Brebner, not included in the firm's title, contributed significantly as a partner from 1878–90 (**2**, 115,117). David Alan Stevenson also joined this partnership in 1878 and Charles Stevenson in 1886. From 1811 the office was at 1 Baxter's Place, Edinburgh; from c.1832 it was above the Northern Lighthouse Board's offices at 84 George Street, Edinburgh; and from 1936, nearby at 90a George Street.

NORTHERN LIGHTHOUSE BOARD ENGINEERS (1787–2010)

Thomas Smith	22 January 1787 – 11 July 1808
	(salaried from 26 December 1793)
Robert Stevenson*	12 July 1808 – 14 December 1842
Alan Stevenson	14 January 1843 – February 1853
David Stevenson	16 March 1853 – March 1855?
D. & T. Stevenson	1 January 1855 – 25 June 1884
Thomas Stevenson	25 June 1884 – 7 March 1885
T. & D. [A.] Stevenson	7 March 1885 – 8 May 1887
David A. Stevenson	2 November 1887 – 31 March 1938
John Oswald (d. 1946)	11 May 1938 – 1 July 1946
John D. Gardner (d. 1964)	31 July 1946 – 2 July 1955
Peter H. Hyslop	1 August 1955 – 30 June 1978
John H. K. Williamson	1 July 1978 – 30 March 1987
William Paterson	1 April 1987 – 20 July 2000
Moray Waddell	3 July 2000 – still in post

* In July 1797, at Smith's request, the Board agreed to Stevenson acting in his place for the annual visit to the lighthouses.

N.B. The information in the above lists are from various sources and indicative only. They have not been entirely checked against formal documents.

Part II

A new look at the creation of Bell Rock Lighthouse from little-known records

Comprising:

An introduction by the editor

Robert Stevenson's *Account of the Bell Rock Lighthouse* published in 1813

John Rennie's unpublished report of October 1809 to the Commissioners

Appendix A. John Rennie's letter of 12 March 1814 to Matthew Boulton regarding his and Stevenson's roles

Appendix B. Extract from Rennie's letter of 13 February 1807 to Stevenson specifying the curve of the base

AN INTRODUCTION BY THE EDITOR

The contents of the following little-known records complement my understanding of the relative contributions of John Rennie and Robert Stevenson to the creation of the Bell Rock Lighthouse set out in *Bright Lights: the Stevenson Engineers* (1999) (**2**, 35–37).

The first record is an *Account* of the lighthouse written in the year following its completion by Robert Stevenson which, now appropriately illustrated, mostly from his own later work, makes fascinating reading. His authoritative descriptions range from the medieval tolling bell which gave the rock its name, via shipwrecks in 1799 giving impetus to early beacon attempts and his own proposals of 1800, to the as-built lighthouse and its erection, occupation and key participants. He acknowledges then, that to himself and Rennie 'was committed the execution of this great undertaking'.

This acknowledgement in 1812 is of interest in the context of the controversy which arose later between the Rennies and Stevensons as to whose forebear *designed and built* the lighthouse. Although Stevenson does not specify their relative contributions here, perhaps if he had the controversy would never have arisen, but fortunately for clarification in this context the Northern Lighthouse Board minutes confirm Rennie as 'Chief Engineer' and Stevenson as 'assistant engineer to execute the work under his superintendence' [NAS: NLC2/1/1]. It is also clear, mainly from evidence in the Rennie papers [as much of the relevant correspondence has been removed at some time from the Stevenson archive], that from 1806–10, by means of meetings, reports and considerable correspondence, they acted competently and in a friendly manner in the best tradition of the chief engineer/resident engineer relationship, both engaging in design and execution.

Stevenson himself does not seem to have claimed to have been the sole designer of the as-built lighthouse in any of his publications but, as he did not clarify Rennie's fundamental part in this, and as Rennie himself never managed to publish an account [although working on one in 1820 – he died in 1821], the merit for the lighthouse's design and construction became in time generally attributed to Stevenson. Rennie had become aware of how the situation in this respect was developing by 1814 when he wrote: 'I have no doubt the whole merit of the Bell Rock Lighthouse will, if it has not already been, [be] assumed by [Stevenson] … The original plans were made by me and the work visited from time to time by me during its progress.' (**11**, Appendix A).

The second record, now published for the first time, is Rennie's key progress report of 1809 on the project. It is almost unknown because it was omitted by Stevenson from the appendix of reports in his otherwise largely definitive *magnum opus* on the lighthouse published in 1824 (**3**). This omission, and the fact that Rennie's contribution was largely ignored elsewhere in the book, supports Rennie's, and Clerk of Works David Logan's, belief that Stevenson was 'endeavouring to appropriate the whole merit' of the lighthouse undertaking. Its omission from the main reference work on the project undoubtedly deprived readers of the opportunity of appreciating Rennie in overall charge of operations when the masonry of the tower was above the height of the greatest danger of demolition by heavy seas, giving directions, and noting with satisfaction the beneficial effect of his tower profile against wave action. The omission also gave the false impression that Rennie never inspected the work after the building of the tower had barely started.

A comparison of the Stevenson and as-built designs (pages 96 and 102) indicate the considerable extent of Rennie's overall influence on the work as implemented.

His 'new plan' of cycloidal curves for a more slender as-built tower [Appendix B], a significant improvement on the shape adopted at Eddystone Lighthouse, sent from London in February 1807, enabled the detailed working drawings to be progressed under Stevenson's direction at Arbroath by draughtsman David Logan. These drawings, which included many Stevenson design elements in detail, were sent to Rennie for his approval from September 1807 onwards. As part of his overall direction Rennie also furnished holograph sketches of work to be carried out and, in one instance noticed [see page 77], rejected undovetailed masonry prepared under Stevenson's superintendence.

These above-mentioned actions, now confirmed from letters in the Rennie archive, contradict the contention of Robert Stevenson's sons, which misled many later students of this subject, including Robert Louis Stevenson who wrote in 1893 'that Rennie did not design [or] execute' the lighthouse 'and was not paid for it.' (*Records of a Family of Engineers …* 1912, 94). Posterity was led to believe that Rennie's role was solely: 'advising their father in cases of emergency and as having suggested alterations of Stevenson design to which he did not acquiesce (**1**), that Stevenson alone was in complete control of the design, and that on no occasion did Rennie give instructions or directions.' (**7**, 301)

The merit for the most difficult part of the undertaking, its execution versus the sea by means of what were truly innovative expedients dictated by exceptionally difficult site circumstances, was undoubtedly due to Stevenson as the resident engineer. But, Rennie's contribution was a fundamental element in its sustainability, and without his masterly touch on this, the materials used, and his insistence on closer adherence to Smeaton's Eddystone Lighthouse practice, the lighthouse might have shared the fate of Stevenson's 40 feet high Carr Rock stone tower beacon (1813–17) 13 miles SSW of Bell Rock. Most of this tower was swept away by the sea in a storm soon after its completion in 1817, to be replaced, in 1821, by the present six-pillar cast iron, beacon founded on the surviving courses. (**3**, 56)

This review concludes that, whilst recognising Robert Stevenson's exceptional contribution to the creation of this two-centuries-old marvel of lighthouse engineering, no one person was solely responsible for its design and execution. In engineering terms, the as-built lighthouse was essentially a masterpiece of state-of-the-art practice, modelled on Smeaton's hard-won experience at Eddystone. It was designed and executed jointly in the capacities of their Northern Lighthouse Board appointments by Rennie and Stevenson and their dedicated and competent workforce, headed by those named in Stevenson's 1812 account (page 69), not least, the Logans, and the ingenious Francis Watt.

Roland Paxton

ACCOUNT OF THE BELL ROCK LIGHTHOUSE

[by Robert Stevenson]. Published anonymously as Appendix D in James Headrick's *General view of the Agriculture of Forfarshire* (1813), and now definitely attributed by the editor to Stevenson from his letters [NLS: Acc. 10706/2, 17 & 10706/8, 2]. It was written at the request in April 1811 of Sir John Sinclair. In November 1812, 12 copies of an 8-page printing of the *Account* were sent by the printer to Stevenson, who distributed them to Francis Watt and others named at the end but not, seemingly, John Rennie.]

The shores of this county [Forfarshire] are the most contiguous to the reef of rocks well known by the general name of the Cape or Bell Rock, and long so much dreaded by seamen, as to have formed one of the greatest bars to the navigation of the east coast of Great Britain. The want of some distinguishing mark to shew the place of this rock, when overflowed by the tide, was most severely felt for ages; and every philanthropic mind must rejoice that this want is so happily removed, by the erection of a lighthouse, similar to that erected about 50 years since, upon the Eddystone Rocks, in Plymouth Sound, by the celebrated Mr Smeaton [completed in 1759].

The erection of the Bell Rock Lighthouse being evidently a great improvement upon the navigation of the North Seas, and ultimately tending to the commercial advancement of the county of Forfar, in rendering more safe the approach to the harbours of Dundee, Arbroath, and Montrose, some account of the progress and completion of such an undertaking may be expected in a Report of this nature. The facts stated shall be quite authentic; but a very minute detail seems unnecessary, because there is every reason to expect the speedy publication, by Mr Robert Stevenson, engineer, of a particular account of the whole operations, illustrated by engravings [in the event it was to take 12 years to produce! (**3**)]. Such a work must be interesting, not only to the man of science, but to the general reader; and it is pleasant to learn, that the Commissioners for Northern Lights, have liberally encouraged such a publication [by indicating their willingness to contribute 'say £400 towards the cost' – NLS: MS. 19806, 143].

The Cape or Bell Rock lies about 11 miles south-west from the Red Head, a remarkable promontory on this coast, which it resembles in colour and nature, being red sandstone, of a fine grit. As seen at low water of spring-tides, it extends about 2000 feet in length, and about 230 feet in breadth. The north-east, or highest part, on which the lighthouse is built, is only partially dry at low water of neap-tides; but in spring-tides, this part of the rock appears from four to six feet above the water; while at high water of the same tides, it is about 12 feet under water. The surface of the rock is very rugged, and it is with difficulty that one can walk upon it. The lower parts are covered with sea weeds of the larger sorts, and the higher parts with mussels

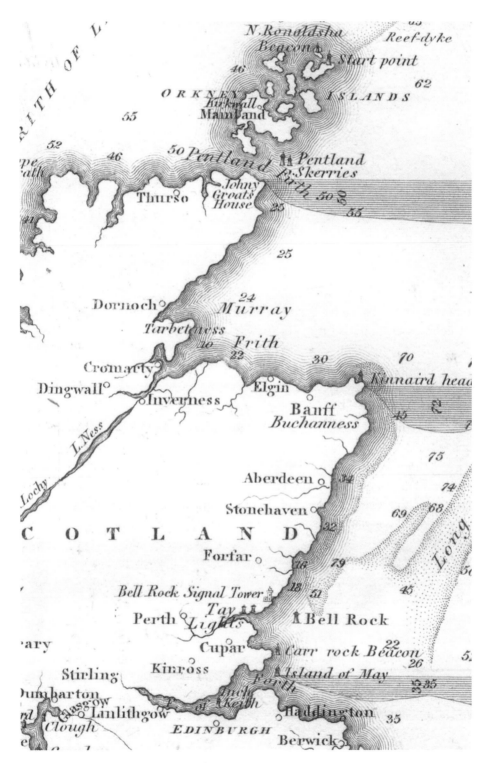

Part of a chart of the east coast of Scotland showing the Bell Rock (3, pl. III)

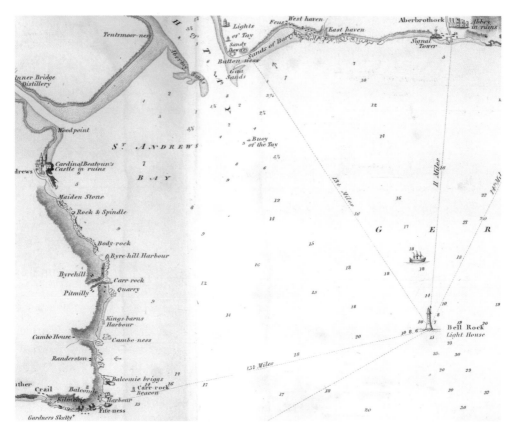

Chart of coast line adjoining the Bell Rock Lighthouse and Carr Rock Beacon. (3, pl.IV)

and whelks, and such kinds of crustaceous and testaceous animals as are common to the shores of this county; and around the rock are caught in plenty the red-ware cod, with other common fishes of these seas.

The most partial examination of any sea chart of this coast, will shew the centrical position which this most dangerous reef unluckily occupies, in relation to all vessels bound, either over seas, or coastwise to or from the Friths of Forth and Tay. Much as the want of some distinguishing mark upon this rock may have been felt during the earlier period of our history, yet while trade was in its infancy, the difficult and expensive nature of such a work may easily be imagined to have been sufficient obstacles to such an undertaking. If we may believe tradition, the pious inhabitants of the Monastery of Aberbrothwick, more than four centuries ago, caused a large bell to be placed upon the rock, so hung that the motion of the waves set it a-ringing, and the mariner was in this manner forewarned of his danger, which is said to have given rise to the name Bell Rock.

In the month of December 1799, the east coast of Great Britain was visited by a dreadful storm from S.S.E., when about 70 vessels, with many of their crews, were

Stevenson's cast–iron Bell Rock Lighthouse proposal of 1800, which he abandoned as impracticable after first examining the Rock in October 1800. It was influenced by Whiteside's Smalls Lighthouse 1776 (7, 128). In 1800 Stevenson was at the outset of his civil engineering career (see page 6).
(3, pl. VII)

lost on these shores, between Fifeness and Aberdeenshire, which created a sensation of the deepest regret throughout the kingdom, and turned the attention of several enterprising individuals towards the Bell Rock; the shipwrecks being attributed to the fear of this fatal reef, hindering seamen from taking shelter in the Frith of Forth, for which the wind was favourable during the storm. Of these, Captain Brodie of the Royal Navy prepared several very ingenious models for the erection of a lighthouse of cast-iron; and went so far as to construct a beacon, consisting of four spars of timber, which stood for several months upon the rock [In 1803 (3, 88)].

The late Mr. Downie, author of the Marine Survey of the east coast of Scotland, proposed the erection of a lighthouse, to stand upon pillars of stone. Mr. Roberts, a well informed sailing master of the Navy, and a native of this county, was also at considerable pains in forwarding the inquiries into this subject. About 25 years since, the unprotected condition of the coast of Scotland, with regard to lighthouses, having been urged in Parliament by a gentleman of this county, George Dempster, Esq; of Dunichen, too well known for his spirited exertion in public affairs, to require any eulogium in this place; in 1786, a bill was brought forward, and certain Commissioners appointed for erecting lighthouses in the northern parts of Great Britain, several of which have accordingly been erected; and no sooner were the funds of that Honourable Board in a promising state, than they projected the great undertaking of a building upon the Bell Rock, to answer the purpose of a beacon by day, and a lighthouse by night.

A bill was accordingly brought into Parliament in the year 1803, which, however, was lost in the House of Lords. The subject was again resumed by the Commissioners in 1806 and a bill was brought forward by the Honourable H. Erskine, then Lord Advocate of Scotland, and seconded by the exertions of the Right Honourable President of the Board of Agriculture, Sir John Sinclair. By this bill, the Commissioners, the better to enable them to erect and maintain a lighthouse upon the Bell Rock, were allowed to extend the collection of the duty for the Northern Lights, to all vessels sailing to or from any port between Peterhead on the north, and Berwick-upon-Tweed on the south, being at the rate of three halfpence per ton upon British, and three pence per ton upon foreign bottoms. The same act authorised the Commissioners to borrow £25,000 from the three per cent. consols, which, with £20,000 of surplus light-duties invested in the funds, made up the disposable sum of £45,000 to proceed with the work.

(The Civil Engineer and Architect's Journal, 1849) Stevenson's undovetailed tower proposal of 1800–06, estimated at £42,635, sent to Rennie 28 December 1805 for consideration [NLS: MS. 19806/1], compared with the as–built design from 1807 under Rennie's overall direction. He insisted on closer adherence to Eddystone's design because of its proven stability' (3, 451). To minimise heavy sea effects he adopted a shape about 20 per cent more slender 30ft above the rock than Stevenson's, with cycloidal curves and lateral dovetailing. The outside stair was omitted. The shape and narrower walls allowed larger rooms. Within these parameters Stevenson had considerable autonomy in detailed design.

Several plans and models had been submitted to the consideration of the Commissioners; but those of their engineer, Mr Stevenson, were ultimately approved of. This gentleman, in the year 1800, made a particular survey of the Bell Rock; and his report [Dated 28 December 1800 after showing models of both designs to Rennie, Professors Playfair and Robison, and Capt. Huddart – all of whom preferred a stone lighthouse. (3, 445)] was afterwards published, along with a letter from the Honourable Captain, now Admiral Cochrane, who, so early as 1793, called the attention of the Commissioners to this important subject. But so various were the opinions of the public regarding even the practicability of the work, and still more concerning the construction of the building best adapted to the situation, that, where so large a sum of public money was necessarily to be expended, the Commissioners judged it proper to submit the subject to the opinion of Mr Rennie [in 1805, preparatory to making the second application to parliament for an Act to erect the lighthouse]. This eminent engineer coincided with Mr Stevenson in thinking, that a building, upon the principles of the Eddystone Lighthouse, was both practicable and advisable at the Bell Rock. [The differences readily apparent from a comparison of the tower designs indicate where the coincidence between the engineers ended]; and to these gentlemen was committed the execution of this great undertaking.

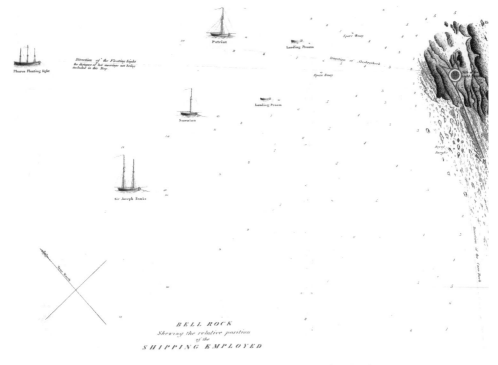

Shipping employed during construction (3, pl. V)

The first object was to moor a vessel as near the Bell Rock as she could ride with any degree of safety, to answer the purpose of a floating-light, and a store-ship for lodging the workmen employed at the rock [The Pharos Floating Light]. This vessel measured 80 tons. She had three masts, on each of which a large lantern was suspended, with lights, which distinguished this light from the double and single lights on the coast. Under the deck, she was entirely fitted up for the accommodation of the seamen and artificers, with holds for provisions and necessaries. Thus furnished, she was moored about two miles from the rock, in a north-east direction, in 22 fathoms water, with a very heavy cast-iron anchor, resembling a mushroom, and a malleable iron chain, to which the ship was attached by a very strong cable. In this situation, the Floating-Light was moored in the month of July 1807, and remained during the whole time the house was building, and until the light was exhibited in February 1811 when she was removed.

The bill for the erection of the lighthouse passed late in the session of 1806 [Act 46. Geo. III. c. 132, 21 July 1806. Both Stevenson and Rennie were examined by a parliamentary committee] and during the following winter, the necessary steps were taken, to have everything in readiness to commence the operations at the rock at the proper season.

A work-yard, upon a lease of *seven years*, was provided at Arbroath, where shades [sheds] for hewing the stones, and barracks for lodging the artificers, when they landed from the rock, were erected. Vessels for conveying the stones from the quarries to the work-yard, and from thence to the rock, were hired or built; and in a few months, Arbroath, always a scene of business and activity, became now the resort of the curious, as well strangers from a distance, as people from the neighbouring towns and parishes of the county, who came to see the preparations for the lighthouse.

Early in the month of August 1807, the operations at the rock commenced, but little was got done towards preparing the rock for the site of the building, till the year following, the chief object of this season's work being to get some temporary erection on the rock, to fly to, in case of an accident befalling any of the attending boats. As the rock was accessible only at low water of spring tides, and as three hours was considered a good tide's work, it became necessary to embrace every opportunity of favourable weather, both under night, by the help of torch-light, and upon Sundays; for the water had no sooner begun to cover the rock, than all the men collected their tools, and went into the boats, which often, with the utmost difficulty, were rowed to the Floating-light.

By such exertions, this work was only to be overcome; and by the latter end of October, the work for the season was brought to a close, after erecting a beacon, which consisted of twelve beams of wood, forming a common base of 30 feet, with 50 feet of height; the whole being strongly held to the rock by bats, and chains of iron.

*Forging beacon fixings and pumping foundation pit, June 1808 (**3**. pl. XI)*

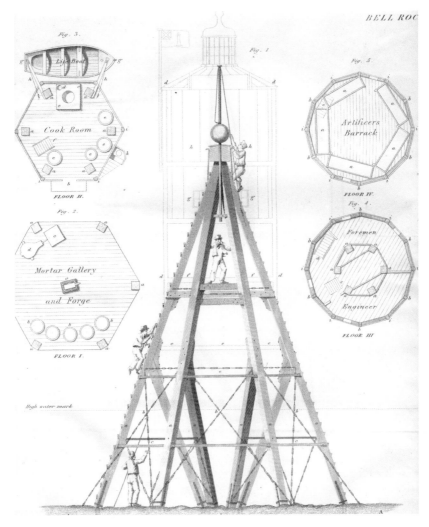

*Temporary beacon 1807 before conversion into a barrack in 1808 (**3**, pl. VIII)*

This beacon, or temporary house, was used as a barrack for the artificers while the work was in progress, and remained on the rock till the summer of 1812, when it was removed.

To the erection of this beacon, the rapidity with which the lighthouse was got up, is chiefly to be ascribed [Although Stevenson conceived and directed the implementation of this remarkable temporary work it is evident from his letters [NLS: MS. Acc. 10706/3, 21, 216] and David Logan's [NLS: MS. 19806, 147] that the design and erection of the beacon, innovative beam and balance cranes and railway were the work of Francis Watt, foreman millwright, employed from 1807–10. In his *Account* (**3**, 496) Stevenson acknowledged that he had 'often profited by Watt's ingenuity'] and it is extremely doubtful if ever it would have been accomplished, without some such expedient, certainly not without the loss of many lives; for in a work of this nature, continued for a series of years, it is wonderful that only one life was lost on the rock, by a fall from a rope-ladder when the sea ran high [Charles Henderson, a smith, 16 October 1810 (**3**, 391)], and another at the mooring-buoys, by the upsetting of a boat [James Scott, a seaman on the Smeaton 21 September 1808 (**3**, 353)].

The operations of the second season were begun at as early a period as the weather would permit, when the preparation of the rock was proceeded with. The risk, and often excessive fatigue, which occurred every tide, in rowing the boats to and from the rock to the Floating-light, made it necessary to have a vessel, which,

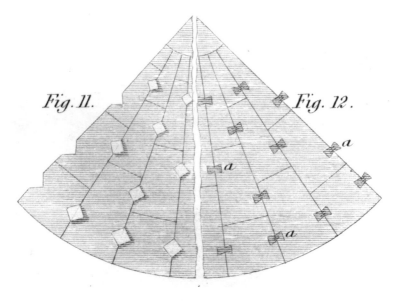

(3, pl. VII)

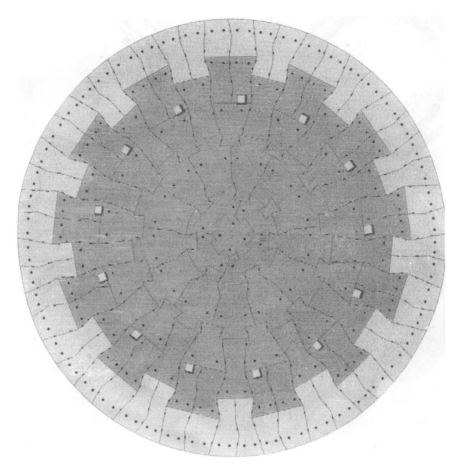

Opposite page: Stevenson's proposed undovetailed Bell Rock Lighthouse courses 1800-06 secured by stone joggles or iron bats (a), compared with [above:] the first entire masonry course as built at Rennie's insistence with dovetailing. (3, pl. XIII)

in blowing weather, could be loosened from her moorings at pleasure, and brought to the lee-side of the rock, where she might take the artificers and attending boats on board. A vessel of 80 tons was accordingly provided, and named The Sir Joseph Banks, in compliment to that worthy baronet, who, ever ready in the cause of public improvement, had lent his aid in procuring the loan from Government for carrying this work into execution.

Through much perseverance and hard struggling with the elements, both during day and night tides, the site of the lighthouse was got to a level, and cut sufficiently deep into the rock. Part of the cast-iron rail-ways, for conveying the stones along the rock, were also got ready: so that on Sunday 10th July 1808, the foundation stone was laid [By Stevenson, in the presence of leading foremen Peter Logan and Francis

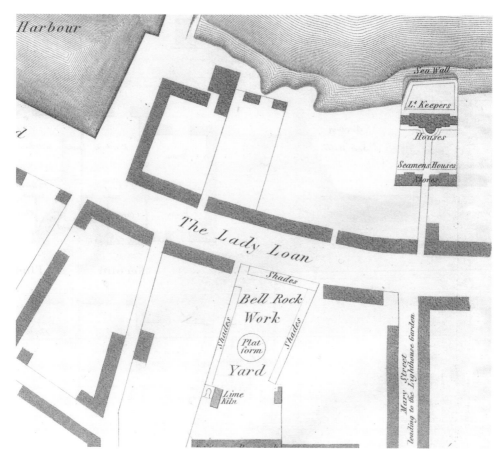

Lighthouse work-yard and shore station, Arbroath. (**3**, *pl. XII*)

James Craw and 'Bassey' transporting stone to and from the work-yard. (**3**, *pl. X*)

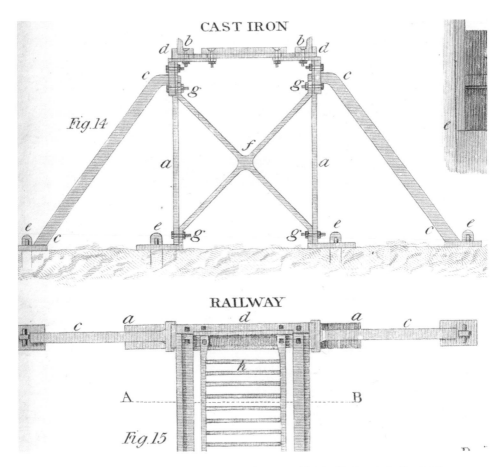

Railway designed by Watt in 1808 – side stays added by Slight 1819. (3, pl. X)

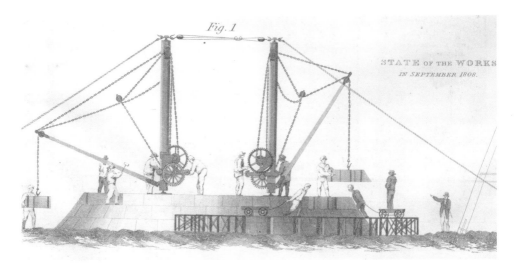

The innovative beam cranes and railway in operation in 1808.(3, pl. IX)

Watt, who applied square, level and mallet to the stone with the benediction 'May the Great Architect of the Universe complete and bless this building.'(**3**, 238). Two days later he formally took over from Smith as Engineer to the Northern Lighthouse Board (**2**, 20)]; and by the latter end of September, the building operations were brought to a conclusion for the season, the first four courses of the lighthouse having been completed.

A stock of materials being procured from the granite quarries of Aberdeenshire, for an outside casing to the height of 30 feet, and from the [Kingoodie] freestone quarries of Mylnfield, near Dundee, for the inside and upper walls, a great number of masons were kept in the work-yard at Arbroath, and every preparation made during the winter months for the work at the rock against next season. [At the work-yard, as an evening occupation to benefit some of the men, no doubt encouraged by Stevenson who paid great attention to training, architectural drawing was taught by David Logan, Clerk of Works – engineering assistant – who in later life became the Chief Engineer of the Clyde Navigation. Logan prepared the working drawings for the as–built tower, including one for each course (**3**, pl. XIII & XVI)].

The stones were wrought with great accuracy, and laid upon a platform, course by course, and there numbered and marked as they were to lie in the building, when

*Off-loading stone into a praam-boat for taking to lighthouse – June 1810. (**3**, pl.XI)*

they were laid aside as ready for shipping for the rock;– a part of the work which was performed with wonderful dexterity; for the vessels which carried them away, were generally dispatched with their cargoes on the tide following that of their arrival.

At the commencement of the operations in April 1809, the four courses built last season were found to be quite entire, not having sustained the smallest injury from the storms of winter. In the arrangements for the work, the first thing to be done was to place the moorings for the various vessels, and stone boats employed in attending the rock, and landing the materials. The machinery for receiving the stones from the praam-boats was erected, and cranes for laying the stones in their places upon the building. With an apparatus thus appointed, the lighthouse was got to the height of thirty feet by the month of September 1809, when the work was again left off during the winter months. Early in the spring of the year 1810, the building was again resumed, but with very faint hopes of bringing the whole to a close in the course of this year: however, as it fortunately happened, not a single stone was lost or damaged, and, by the month of December [On 2nd September 1810, Stevenson laid the last stone, the upper step of the stair – the wooden bridge to the beacon was removed. (**3**, 388)], everything was got into its place; and the interior having soon after been finished, the light was exhibited, for the first time, on the night of the 1st February 1811.

Having now, in a very general way, noticed the various stages of the erection of the Bell Rock Lighthouse, it only remains to give some of its dimensions, and a few other particulars.

The foundation-stone of the lighthouse is nearly on a level with low water of ordinary spring-tides, and consequently the lower part of the building will be about 15 feet immersed in water when the tide has flowed to its usual height at new and full moon. But during the progress of the work, the sea spray has been observed to rise upon the building to the height of 80 feet; and upon one occasion to 90 feet, even in the month of July. The house is of a circular form, measuring 42 feet diameter at the base, from which it diminishes as it rises, and only measures 13 feet at the top, where the light-room rests: Including which, it measures in height altogether 115 feet. To the height of 30 feet it is entirely solid, excepting a drop-hole for the weight of the machinery which moves the reflectors, which hole is only ten inches diameter. The ascent to the door, which is placed at the top of the solid, is by a kind of rope-ladder. A narrow passage leads from the door to the stair-case, where the walls are seven feet in thickness: at the top of the stair-case, which is 13 feet in height, the walls get thinner, and diminish gradually to the top.

Above the stair-case, the ascent to the different apartments is by means of wooden-ladders; and the remaining 57 feet of masonry is divided by five stone floors into rooms for the light-keepers, and stores for the light, and the light-room is placed on the top of the building. The three lower apartments have each two small windows, while the upper rooms have each four windows; and the whole are provided with strong shutters, to defend the glass against the sea in storms.

The two first courses of the building are entirely sunk into the rock; and the stones of all the courses are dove-tailed and let into each other, in such a manner as that each course of the building forms one connected mass, and the several courses are attached to each other by joggels of stone and oaken trenails, upon the plan of the Eddystone Lighthouse formerly alluded to. The cement used at the Bell Rock was a mixture of pozzolana earth, sand and lime; which last was brought from Aberthaw in Wales, where the lime for the Eddystone Lighthouse was got.

Round the balcony of the light-room, there is a cast-iron rail, curiously wrought like net-work, which rests on bats of brass. The light-room is 12 feet diameter, and 15 feet in height, made chiefly of cast-iron, with a copper roof. The windows are glazed with large plates of polished glass, which is one quarter of an inch thick.

The light is from oil, with Argand burners, placed before silver-plated reflectors, hollowed out to the parabolic curve. That the Bell Rock light may be distinguished from all others on the coast, the reflectors are ranged upon a frame which is made to revolve upon a perpendicular axis once in three minutes. Before some of the reflectors are placed shades of red coloured glass; so that the effect produced in each revolution of the frame with the reflectors, is a light of the natural appearance, and a light with the rays tinged red, with intervals of darkness between the lights. In a favourable

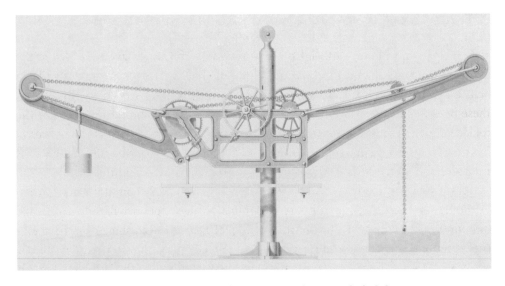

Logan's drawing of Watt's iron balance crane used to erect the lighthouse tower.
© *NLS: MS. Acc. 10706 (for a published version with plan see* **2**, *39)*

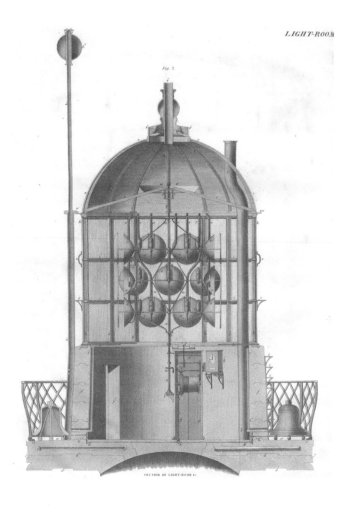

Light-room, with revolving array of silvered copper reflector lamps, signal ball, fog bells and ornamental iron balcony railing.
(**3**, *pl. XX*)

state of the atmosphere, these lights are so very powerful, as to have been seen at the distance of 25 miles. During the continuance of thick and foggy weather, two large bells are tolled night and day, by the same machinery which moves the lights; and as these bells may be heard in moderate weather considerably beyond the limits of the rock, the mariner may be advertised of his situation, in time to put about his vessel before any accident can happen; for in thick and hazy weather, she might otherwise run ashore upon the rock, notwithstanding the erection of the lighthouse.

About the commencement of these works, it was a very common saying, that: 'Although the Bell Rock Lighthouse were built, (which it never will be), no one will be found hardy enough to live in it.' The sequel has, however, shewn the fallacy of such a supposition; for no sooner was the house ready for possession, than numerous applications were made for the situation and many were of course disappointed. Of these applicants, a principal light-keeper and three others were nominated, and took up their abode at the term of Martinmas 1810 [November 11], and each in his turn gets ashore on liberty at the end of every six weeks, and remains a fortnight, when he goes off to the lighthouse again.

The pay of the light-keepers is about £50 *per annum*, with provisions while they are at the lighthouse; but ashore they provide themselves. At Arbroath, there are buildings erected, where each keeper has apartments for the accommodation of his family; and, connected with this establishment, there is a very handsome signal-tower, 50 feet in height, in which an excellent telescope is kept, and signals arranged with the people at the rock for the attending vessel; this vessel is about 40 register tons, and is capable of carrying a large enough boat for landing at the rock in moderate weather, with stores, provisions, fuel and water; and the master of this vessel has also the charge of the stores at Arbroath.

This establishment, which is as complete as can well be imagined, says much for the humane consideration and proper liberality of the Honourable Board of Commissioners. At present, the exact amount of the expence of the erection of the lighthouse, and establishment connected with it, cannot be ascertained, but will probably be about £55,000 Sterling. Whether, therefore, we look to the peculiar position of the reef on which this lighthouse is built, or to the success which has attended the operations, from their commencement in 1807, to the period of their final conclusion in December 1810, this work will be found to do equal honour to the spirit and resources of the age in which we live.

As many of the seamen and artificers engaged in this memorable work claim kindred to the county of Forfar, besides those brought from Mid-Lothian, Aberdeenshire, &c. &c. it may be proper to mention, much to their honour, that the Magistrates of Arbroath give the most ample testimony of their orderly conduct during the three years which the work was going forward; and having been engaged in the erection of the Bell Rock Lighthouse, will always be a sufficient passport for abilities in the line of their profession. Although in a work of such extent, necessarily divided into various departments, it would be impossible to mention all who signalised themselves for their faithful exertions, yet we cannot withhold the mention of the following gentlemen, upon the best authority:

> Mr Peter Logan, foreman to the building operations at the rock
> Mr David Logan, draughtsman, and foreman at the work-yard.
> Arbroath [Also designated 'Clerk of Works' by Stevenson (**3**, p.490)]
> Mr Francis Watt, foreman of the joiners
> [Also designated 'foreman mill-wright' by Stevenson who, on 26 November
>   1812 went to some trouble to get a copy of the separate printing of this
>   *Account* to Watt, "I will carry [one] to Arbroath with me tomorrow and give
>   to Captain Brown who occasionally tells me of your welfare and who will
>   know your address in London" [NLS: Acc. 10706/8, 2].

Mr James Dove, foreman of the smiths. [d. 1843]

Mr James Slight, mould-maker. [1786–1854, became a civil engineer]

Captain James Wilson of the Floating-light, and landing-
    master at the rock.

Captain David Taylor of the Sir Joseph Banks tender. [1768–1843]

Captain Robert Pool of the Smeaton stone-lighter.

Captain James Spink of the Patriot stone-lighter.

Captain John Reid, acting-master, and principal light-
    keeper of the Floating-light. [1753–1843]

And Mr Lachlan Kennedy, clerk and cashier, Engineer's Office.

Several of these gentlemen are still in the Lighthouse service; while others have removed to works of celebrity, or are engaged in business on their own account.

It must be a matter of very general satisfaction to learn, that the lighthouse, in its entire state, has sustained no injury whatever from the storms of the first and second winters. And we shall conclude these remarks with observing, that Mr [John] Forrest, the Superintendant for the instruction of the light-keepers, after remaining at the lighthouse from December 1810 till April 1811, reported, that '*the lighthouse was as dry and comfortable as any house in Edinburgh.*'

Bell Rock Lighthouse kitchen in 1865, as used by R.M. Ballantyne when writing his story of the great struggle between man and the sea to create the lighthouse (9). To get a feel for his subject he lived in the finely appointed 'Stranger's Room' or library under the light-room for a fortnight. Note the dovetailed cantilever ends in the floor insisted on by Rennie.

*Above: The library
or 'Strangers Room'
from a painting by
R.M. Ballantyne, who
occupied the room in
1865 when writing*
The Lighthouse.
*Complete with Turkey
carpet, the marble
Stevenson bust and
tablet, pedestal table
with bible (prayers
were regularly read
here), open bookcase and
other fine fittings of 'oak
timber executed in Mr.
Trotter of Edinburgh's
best style … [and]
handsomely decorated
panel–work painted
by Mr. Macdonald
of Arbroath'. It is
understood that the
room remained more or
less like this until the
1960s.
© NLS: Acc. 11962*

*Below: Bell Rock
Lighthouse during
a gale (***6***)*

Logan's section of the completed lighthouse with its 90 courses and 'War without and peace within' – R. M. Ballantyne's artistic frontispiece of The Lighthouse *compared, to show its internal arrangements – 1811–1865. (3, pl.XVI) (9)*

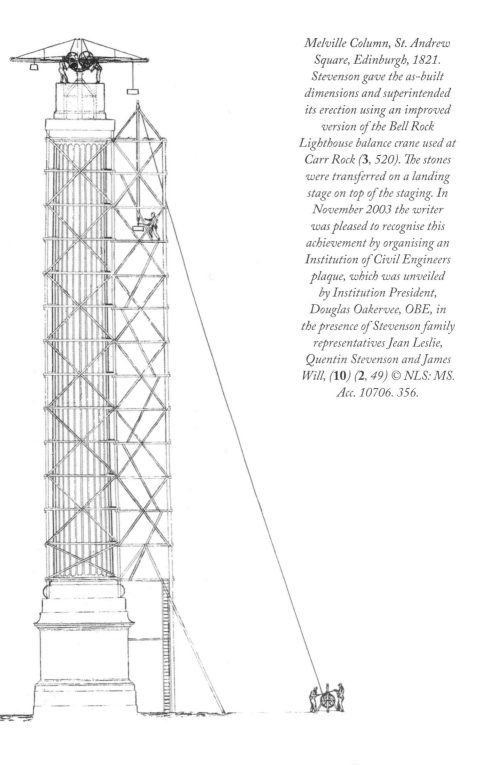

Melville Column, St. Andrew Square, Edinburgh, 1821. Stevenson gave the as-built dimensions and superintended its erection using an improved version of the Bell Rock Lighthouse balance crane used at Carr Rock (3, 520). The stones were transferred on a landing stage on top of the staging. In November 2003 the writer was pleased to recognise this achievement by organising an Institution of Civil Engineers plaque, which was unveiled by Institution President, Douglas Oakervee, OBE, in the presence of Stevenson family representatives Jean Leslie, Quentin Stevenson and James Will, (10) (2, 49) © NLS: MS. Acc. 10706. 356.

Bell Rock Lighthouse

Report by John Rennie to the Commissioners of Northern Lights &c.,
Edinburgh 2 October 1809.

[Omitted from Appendix IV – Reports of Rennie and Stevenson –
in Stevenson's *Account* of 1824 (**3**) and now first published]

Gentlemen,

I went to Arbroath on the 22nd [September] but owing to the badness of the weather, I did not get out to the Bell Rock until the 24th when I was fortunate in having a good ebb and very favourable weather which enabled me to examine the building and the Rock round its foundation with great correctness as also the state of the beacon with the place on the top for the accommodation of the workmen and the iron railways &c. [This statement emphatically contradicts D. Alan Stevenson's conjecture that Rennie only 'saw the tower from a distance about high-water' (**7**, 301)]

The tower for the lighthouse is now raised to the height of about 30 feet. The outside is all of granite except the two upper courses which are of Kingoodie stone [From Mylnfield, on the north side of the River Tay, five miles west of Dundee] & I have much pleasure in saying that the work is well executed and the cement of the best quality no part of which has in any degree failed. While however [in] the courses of stone now laying under the level of the tide it would appear that some small part of the green mortar had been washed out and there is a little weeping of water in those places, but this is very trivial. I have directed it to be examined and filled with grout and if a little oakum is put into them to secure the grout until it has time to harden and the outside is then pointed with Roman Cement there is no doubt that the whole will be perfect. [This again contradicts D. Alan Stevenson's conjecture that Rennie never gave directions (**7**, 301). Rennie, from his considerable experience of hydraulic mortars and contacts, also specified and sourced the Roman Cement and pozzolana earth used for jointing the masonry].

The rock round the foundation of the tower does not appear to waste & some small rub[b]le of cement which had been laid round it the end of last year ... is as perfect as the day it was laid. [It] is true the building was then so low that there could be no heavy action of the sea against it but where the vacancies are filled in with stone as intended I have no doubt that the action of the sea will not produce any material effect on it. The curve of the outside of the tower answers fully to every expectation I had formed of it, the sea plays with ease round it and I trust it will be found when finished the compleatest work of its kind. [Contrary to the conjectures of D. Alan

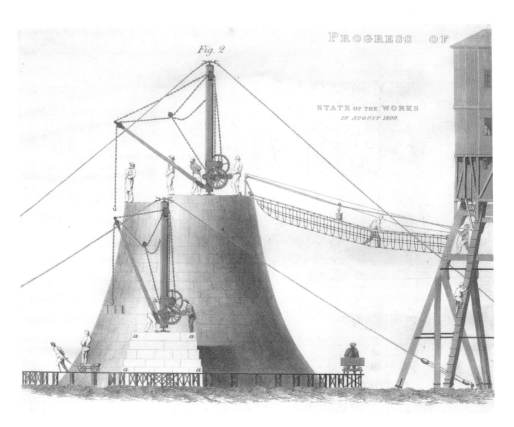

Bell Rock Lighthouse works August 1809. Rennie visited the site on 24 September 1809, and examined the tower at a height of 31 ft 6 in, with about 60 per cent of its masonry laid (3, 471). It was almost certainly past the greatest danger of demolition by heavy seas. This was probably Rennie's last site visit before its completion in 1810. Note the innovative cranes, railway and beacon conceived and directed by Stevenson, designed and erected by Francis Watt, and approved by Rennie. (3, pl. IX)

Stevenson (7, 301). The implementation of the cycloidal curves, slenderer tower and dovetailing of Rennie's plan undoubtedly reduced the risk of storm damage in heavy seas, particularly at or near High Water level during construction of the first 30 ft of the tower.]

The beacon has shook considerably by the swell of the sea that breaks on it not by a regular vibration in the direction of its height but by a kind of twist. Mr. Stevenson has very properly directed side braces to be put to the main braces of it, and all the iron chains and work to be tightened up which I have little doubt will render it secure. This beacon has been a very expensive work, but it has been of great use in facilitating the operations at the rock, and it will be important to secure it in such a manner as to stand for another year which I hope the measures now taking will enable it to do. [Rennie would have met some of the 24 workmen accommodated in the beacon still engaged until November on laterally bracing its supports and extending the railway.]

Mr. Stevenson has thought proper to increase the number of landing places and to lay down iron railways from thence to the tower, these are attended with a heavy expense, but no doubt they will facilitate the conveyance of stone to the tower. I was however in hopes some part of this expense would have been saved [Rennie reluctantly accepted the increase in the number of landing places and railway extension as traditionally the resident engineer had considerable autonomy in the design and construction of temporary works.]

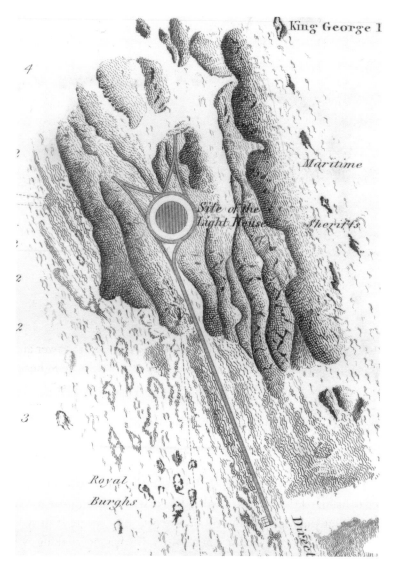

Bell Rock – Plan showing the proposed railway from the landing places to the lighthouse site.
(3, pl. V)

I come now to a subject much less pleasant namely the supply of stone from the Kingoodie Quarries. When the contract was entered into with Mr. Mylne the quarry had a favourable appearance and there was no doubt entertained by any one that a most ample supply of stone for the work might be had from it but unfortunately this has not been the case, the quarry having in a great measure failed in respect to the production of large blocks of stone. From the best information I have been able to procure Mr. Mylne has not been deficient in exertions but the quarry has become so deficient in its supply of large blocks that he has been unable to comply with his contract.

Notwithstanding, every block fit for the works of the Bell Rock have been appropriated to it. On the 25[th September] I examined the quarry [Kingoodie] with Mr. Logan Junr. [David Logan], and consulted with Mr. Mylne's agent and the mason belonging to the Commissioners who was sent to examine the work what was the utmost that could be expected from this quarry by the middle of next month and they were of opinion that not above 200 tons could be produced and as this [frost susceptible] stone does not answer to be quarried in Winter there appears no hope that the men can be kept employed in the Work Yard longer than about Christmas. A necessity therefore exists of getting stone from some other place or discharging at least one half of the masons, a measure which will not only occasion great delay in the completion of the work but incur a great additional expense in the completion of the lighthouse.

The solid part of the tower being compleated the rest of the work should proceed with much greater celerity. There is stone enough at Arbroath to compleat the stair case part which is about 20 feet high, indeed nearly half of this is already built on the platform in the Work Yard and the rest may probably be done by the end of next month. The stone that is expected from Kingoodie will complete two stories more, three stories and the cornice and platform will then be wanted which may probably require five hundred tons of stone, and my advice is that the stone should be got from the quarries of Craigleith which will cost a little more than that from Kingoodie, but as the quality is as good for standing the spray & weather and as the delay of one year will probably incur an additional expense of at least £5,000, I cannot hesitate a moment in recommending this measure to the immediate attention of the Commissioners. [Stevenson acted on Rennie's advice and Craigleith stone was used for the cornice and light-room wall from the 81st course upwards (**3**, 371 – page 71), about 1,700 cubic feet before dressing.] The establishments of vessels, superintendants &c is very heavy and the sooner this is reduced the better as the work will be found sufficiently expensive with all the œconomy that can be practised.

The manner of dovetailing the stones of the solid part of the tower is nearly the same as that of the Eddystone and this plan will be followed to the top of the stair case. I have however to recommend a mode some what different for the hollow part of the surrounding wall of which should be dovetailed in a mode I have already drawn out. [This and other sketches of alterations by Rennie to Stevenson's work were still in David Logan's possession in 1820, including *the courses as undovetailed at the centre which were adopted by Mr. Stevenson but afterwards rejected by Rennie.* [NLS: MS. 19806, 147–150]. In actioning this rejection Stevenson sent Rennie a drawing of one of the *courses at the centre dovetailed to the centre stone for your approbation* [2.9.1807. NLS: MS. 19806, 33.] pointing out that the stone laid aside could be used elsewhere [NLS. MS. 19806, 33]. This, and other letters in the Rennie archive, contradict D. Alan Stevenson's conjecture that Rennie did not produce any drawings other than the tower outline and that the working plans were not submitted for Rennie's approval (**7**, 205)]. Even temporary works were discussed with Rennie and his advice sought [April 1808. NLS: MS. 19806, 65–67].

The stone floors in the Eddystone were formed by an arch in the form of a dome springing from the surrounding walls to strengthen which chain bars were laid into the wall. I propose that these should be done with large stones radiated from a circular block in the middle to which their interior ends are to be dovetailed as well as the radiated joints and then connected to the surrounding walls by means of a circular dowel. By this means the lateral pressure on the walls will be removed, the whole will be connected as one mass and no chain bars will be wanted except under the cornice. [This proposal, a development of Stevenson's 1800–06 design concept but incorporating end dovetailing, was adopted. The ring chain bar referred to was let into a groove in the 81st course which was the springing of the cornice and top apartment dome.] Thus the whole will be like a solid block of stone excavated for the residence of the light keepers, stores, &c.

The outside stone should not be dressed so smooth as it is now doing. As by the use of Craigleith stone I have great reason to believe if the next Summer should prove any way favourable for working on the rock that the light house will be completed. [The mason-work was completed on 2nd September 1810.]

I advise that the lantern should be prepared without loss of time and I beg leave to recommend to the Commissioners the use of coloured glass in the light similar to what has been lately practised in some of the new light houses built by the Trinity House of London. One of them has been seen lately by Mr. Stevenson as well as myself and we consider it a great improvement in the distinguishing of the lights. Whether the red glass used in these lights is the best that can be adopted I will not pretend to say, but this point can easily be ascertained by experiment. [Red glass was used on Stevenson's initiative.]

   I am,

     Gentlemen,

      Yours &c.

       [John Rennie]

The Commissioners
of Northern Lights
&c. &c. &c.

Editor's notes

Transcribed by myself from Rennie's 'Reports' [pp 296–8] at the Institution of Civil Engineers' library in London, which is probably the only surviving text, no copy having been located in Scotland.

A note from the Northern Lighthouse Commissioners' Secretary Charles Cunningham dated 25 December 1810 confirmed that he was ready to settle Rennie's bill for the September/October 1809 visit and report. In all, Rennie is understood to have received £440 from the Northern Lighthouse Board for his Bell Rock Lighthouse work (**7**, 304).

Rennie is known to have made at least four working visits to the Bell Rock from 1805–09. These were on 16 August 1805 [**3**, 447]; 6 October 1807 (inspecting foundation work and the beacon as just completed before its conversion into a barrack the following year) [**3**, 463]; 25 November 1808 (inspecting the beacon and first four courses) [NLS. MS. 19894, 15]; and 26 September 1809 (inspecting the tower masonry about 60 per cent complete and work in progress strengthening the beacon and on extending the railways).

D. Alan Stevenson was aware of the first two visits, but states that Rennie also visited the rock in December 1808 and October 1809 (**7**, 304), which probably refer to the November 1808 and September 1809 visits. If so, this and the other contradictions of Rennie's role already noticed, suggest that D. Alan Stevenson had not seen Rennie's report of 2 October 1809 when writing his book in 1959 (**7**).

Appendix A

John Rennie's letter of 12 March 1814 to Matthew Boulton

[Partly first published in Smiles, S., *Lives of the Engineers: Smeaton and Rennie*. London, 1874, 295, and in subsequent editions]

London March 12th 1814
Private

Dear Sir

I duly received your letter of yesterdays date & shall with much pleasure comply with your request.

Mr Robert Stevenson was bred a Tin Smith & Lamp Lighter in which line he was employed by a Mr Thomas Smith a considerable Manufacturer in that line in Edin[bu]r[gh] & who had the care of the Reflectors & Lamps belonging to the Commissioners of Northern Lights. While in Smith's employment he married His Daughter, & Smith, advancing in years, employed Stevenson to Look after the Northern Lights. This he did for several years, when Smith declined the situation, & Stevenson was elected in his place. When the Bell Rock Light House was erected, Stevenson was employed to Superintend the whole, there being a regular Mason under him & a carpenter. The original plans were made by me, & the work visited from time to time by me during its progress. When this work was Completed Stevenson considered that he had acquired sufficient knowledge to start as a Civil Engineer & in that line he has been most indefatigable in looking after employment, by writing & applying wherever he thought there was a chance of Success. But few weeks passed without a puff or two in the Edin[bu]r[gh] Papers. He has taken the Merit of applying Coloured Glass to Light Houses which he stole from Huddart & I have no doubt the whole Merit of the Bell Rock Lighthouse will, if it has not already been [be] assumed by Him. He has not however been successfull in getting into employment as a Civil Engineer and in Consequence being a partner in the Greenside Co[mpan]y has lately Started a Manufactory of Steam Engines, & I have no doubt the principal if not the sole object of his sending a Man to your Manufactory, is to acquire information respecting them, & to entice away some of your principal Workmen. You ought therefore to be particularly Circumspect in your transactions with this Man & by no means to admit him or any of his people into your Manufactory.

I am much obliged by your enquiries after my health which instead of suffering from the severe weather has rather improved by it. When this weather will end God only knows. We had a heavy fall of snow here yesterday which now lyes thick & the frost is very sharp. I hope this will find Miss Boulton & yourself in perfect good health as also Mrs. Watt and Mr. Ja[me]s W[att] to whom please to make my [best?]

Compliments.
I remain
Dear Sir
truly yours

John Rennie

M. R. Boulton by
Soho

Letter of John Rennie (1761–1821)
to Matthew Robinson Boulton (1770–1842), 12.3.1814. **(11)**
Boulton Papers, Birmingham Central Library
Transcribed by W. T. Johnston: 18.5.2010.

Editor's notes

The context for this private letter was a business enquiry into Stevenson's background from Boulton & Watt who were furnishing the plated metal for the silvered-copper parabolic lamp reflectors to the Greenside Company, who by then had begun making them for the Northern Lighthouse Board. Previously Boulton & Watt had supplied the reflectors ready formed. The Greenside Company of which, according to Rennie, Stevenson was a partner, wished to send a coppersmith to Boulton & Watt's Soho Works to receive instruction in management of the metal and reflector polishing.

The disparaging element in the letter is not the first known instance of this Rennie characteristic. In November 1806 he had been remarkably critical to Stevenson about Telford (in the context of the latter's 'preparatory step for making a design for the lighthouse in 1803 in association with Murdoch Downie'). Rennie wrote: '[Telford] has no originality of thought and has all his life built the little fame he has acquired upon the knowledge of others, which he has generally assumed as his own'. [NLS: MS. Acc. 10706, 73, f. 55]. These unworthy generalisations were probably fuelled by Rennie's resentment of Telford's parliamentary based progress in improving the infrastructure of the Scottish Highlands. In April 1805 Telford had found it necessary to counter Rennie's illiberal treatment of his character in every quarter in a letter to James Watt [Rolt, L.T.C. Thomas Telford , 1958, 149].

Even allowing for some over-reaction by Rennie, he clearly felt strongly by 1814 that Stevenson was beginning to assume the whole merit of the Bell Rock Lighthouse. Rennie's outline of their roles on the project was based on his own extensive experience. This is still typical of the traditional chief engineer/resident engineer relationship, with the former setting and taking responsibility for the main design and execution parameters and the latter preparing the working and temporary works drawings and receiving only occasional site visits from his chief. From some of his comments Stevenson seems to have expected more of Rennie, but this was his first experience of such an arrangement and an engineering project on this scale.

In his comment about the merit of applying coloured glass to lighthouses, Rennie was probably reflecting the view of his friend Capt. Joseph Huddart (1741–1816), a leading authority on marine surveying and harbour engineering. Stevenson certainly developed this concept.

Regarding Rennie's conjecture about the Greenside Company's interest in gaining information on steam engine manufacture it seems unlikely that Stevenson would have wished to be involved in an operation of this technical complexity. In 1825, when invited by David Brewster to write the article 'Steam Engine' for the *Edinburgh Encyclopaedia*, he replied: 'I should be afraid of disappointing you every way'. [NLS: MS. 10706/16]

Rennie's cycloidal curves as implemented to the 42ft diameter base, checked by the editor and Robert Mackay NLB Civil Engineer when the helicopter pad was being constructed in 1986.

Appendix B

Extract from Rennie's letter of 13 February 1807 to Stevenson specifying the base diameter and curvature of the lower part of the lighthouse tower as it was incorporated into the working drawings and implemented. [NLS: MS. Acc. 10706/63, 65.]

> … I have sketched out a new plan for the [Bell Rock] Light house & I trust this will give you satisfaction. The first or lower course of stones forms a circle of 42 feet dia[mete]r, the second course 38 feet, the third 35, the fourth 33 & the fifth 32, forming by these dimensions a cycloidall base which under all circumstances seems the best. I hope the plans* will be ready by the end of next week when they shall be forwarded to you.

* Only one of these drawings (there may have only been one) appears to have survived, dated 21 February 1807 – an outline of the base and tower. Stevenson was anxious to get these details quickly in order to proceed with the working drawings and moulds of undressed stone sizes for the quarriers, and he requested an outline from Rennie on 14 February 1807 [NLS: MS. 19806/19.] Soon after this, as their letters crossed in the post, Stevenson would have received Rennie's letter fixing the base diameter at 42ft and the lower tower curvature. Details of any other Rennie plans and instructions sent about this time may never be known as his letter to Stevenson of 23 February, which possibly accompanied the plans, is missing from its letter-book in the Stevenson archive. The other possible source, Rennie's outgoing letter books for 1807–13 with copies of his letters to Stevenson, is not to be found at the National Library of Scotland or the Institution of Civil Engineers and may no longer exist.

Despite the urgency for David Logan to prepare the working drawings under Stevenson's direction, it was to be about 15 months before the first entire course of 123 stones was assembled on the work-yard platform at Arbroath ready for shipping out to the rock. This was largely because of the difficulty of obtaining granite blocks of the size and quality required from Aberdeen.

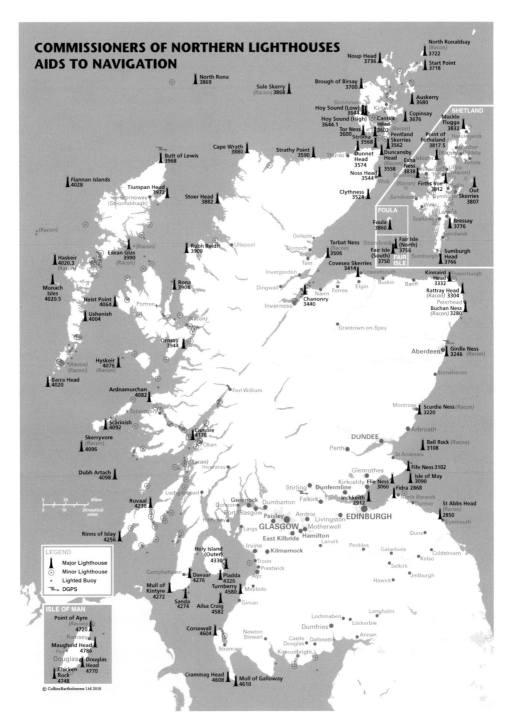

Commissioners of Northern Lighthouses Aids to Navigation 2010
*(updated from **2**, 168) ©CollinsBartholomew Ltd 2010*

Part III

Lighthouses by Smith and the Stevensons –
an illustrated chronology

[updated from the list in Appendix 2 of *Bright Lights: the Stevenson Engineers* (**2**, 191–195) which includes additional detail and grid references]

• *Major Light > 15 miles*
AIS – *Automatic Identification System*
DGPS – *Differential Global Positioning System*
RACON – *RAdar beaCONs*

*The Bell Rock works in August 1810 (***3***, pl. XVIII)*

Thomas Smith

Lighthouse	Established	Automated	Notable Historic Event	Status (2010)
Kinnaird Head • (Fraserburgh)	1787	–	Closed in 1990 New light established in 1990•	Now Museum of Scottish Lighthouses 1990 light Operational
Mull of Kintyre• (Argyll)	1788	1996	Rebuilt 1820's by Robert Stevenson	Operational
North Ronaldsay • (Orkney)	1789	1998	New lighthouse built by Alan Stevenson 1954	Original Beacon Ancient Monument 1954 light Operational RACON Public Access Station (Seasonal) Visitors Centre and privately owned holiday cottages
Tay Lights (Firth of Tay)	Already existing, modernised 1789			Non NLB lights
Eilean Glas • (Scalpay, Harris)	1789	1978	New light built 1820s by Robert Stevenson	Original tower Historic Monument New Tower Operational RACON Cottages privately owned
Pladda • (Arran, Bute)	1790	1990	New light built 1820s by Robert Stevenson	Operational Cottages privately owned
Leith Pier (Midlothian)	Already existing, modernised 1790			Non NLB light
Portpatrick (Wigtownshire)	1790		Discontinued 1900	
Little Cumbrae (Bute)	1793		First 1757 lighthouse tower still exists	Owned/operated by Clyde Port Authority
Pentland Skerries • (Orkney)	1794	1994	New lights built 1830 by Robert Stevenson	Operational Cottages privately owned
Cloch * (Renfrewshire)	1797			Owned/operated by Clyde Port Authority
Inchkeith •* (Fife)	1804	1986		Operational
Start Point •* (Sanday, Orkney)	1806	1962	Rebuilt 1870 by David Stevenson	Operational Public Access Station (Seasonal)

* With Robert Stevenson

Robert Stevenson

Prior to 1808 Robert Stevenson assisted Thomas Smith with Cloch, Inchkeith and Start Point

Lighthouse	Established	Automated	Notable Historic Event	Status (2010)
Bell Rock • (North Sea)	1811	1988	Fire 1987	Operational RACON
Toward Point (Argyll)	1812	1974		Owned/operated Clyde Port Authority
Isle of May • (Fife)	1816	1989	Isle of May site of first Scottish Lighthouse in 1636. Original tower modified circ. 1816, still exists.	Operational Cottages privately owned
Corsewall • (Wigtownshire)	1817	1994		Operational AIS Station Cottages now operate as hotel
Point of Ayre • (Isle of Man)	1818	1991	A second light "The Winkie" was established around 1890, this was discontinued in 2010.	Operational RACON Cottages privately owned
Calf of Man (Isle of Man)	1818			Discontinued 1875 Now owned by Manx National Heritage
Sumburgh Head • (Shetland)	1821	1991	Rebuilt 1914	Operational DGPS Station AIS Station Cottages privately owned
Rinns of Islay • (Argyll)	1825	1998		Operational
Buchan Ness • (Aberdeenshire)	1827	1988		Operational RACON Cottages privately owned
Cape Wrath • (Sutherland)	1828	1998		Operational Cottages privately owned
Tarbat Ness • (Ross & Cromarty)	1830	1985		Operational RACON Cottages privately owned

Name (location)	Year	Notes	Status
Mull of Galloway (Wigtownshire)	1830		Operational Public Access Station (Seasonal)
Dunnet Head (Caithness)	1831		Operational Cottages privately owned
Douglas Head (Isle of Man)	1832	Rebuilt 1859	Operational Cottages privately owned
Girdle Ness (Aberdeenshire)	1833		Operational DGPS Station RACON Cottages privately owned
Barra Head (Barra)	1833		Operational
Lismore (Argyll)	1833		Operational Cottages privately owned

Alan Stevenson

Name (location)	Year	Notes	Status
Little Ross (Kirkcudbright)	1843	Lightkeeper murdered in 1960	Operational Cottages privately owned
Isle of May (Leading light)	1844		Discontinued 1887 when North Carr Lightship became operational
Skerryvore (Argyll)	1844	Fire 1954	Operational RACON
Covesea Skerries (Moray)	1846		Operational
Chanonry Point (Ross & Cromarty)	1846		Operational Cottages privately owned
Cromarty (Ross & Cromarty)	1846		Discontinued 2006
Loch Ryan (Cairn Point)	1847	Cottages demolished	Operational
Noss Head (Caithness)	1849		Operational Cottages privately owned

Lighthouse	Established	Automated	Notable Historic Event	Status (2010)
Ardnamurchan • (Argyll)	1849	1988		Operational Public Access Station (Seasonal) Visitors Centre and privately owned holiday cottages
Sanda • (Kintyre, Argyllshire)	1850	1993		Operational Cottages privately owned
Hoy Sound (High) • (Orkney)	1851	1978		Operational Cottages privately owned
Hoy Sound (Low) • (Orkney)	1851	1966		Operational Cottages privately owned
Arnish Point (Stornoway)	1853	1963	Transferred to Stornoway Harbour 2003	

Thomas Stevenson

Tay South Ferry (Firth of Tay)	1849		Discontinued, still exists	Non NLB light

David and Thomas Stevenson

** Without Thomas Stevenson

Out Skerries •** (Whalsay, Shetland)	1854	1972		Operational Cottages privately owned
Muckle Flugga •** (North Unst, Shetland)	1854	1995	Rebuilt 1857 with Thomas	Operational
Davaar •** (Campbelltown, Argyll)	1854	1983		Operational Cottages privately owned
Ushenish • (South Uist)	1857	1970	Cottages demolished	Operational
Rona • (Inverness-shire)	1857	1975		Operational

Location	Established	Automated	Notes	Status
Kyleakin (Skye)	1857	1960	Gavin Maxwell, author, bought the lighthouse cottages in 1963	Discontinued 1993 on construction of the Skye Bridge. Eilean Bàn Trust ownership in process
Isle of Ornsay • (Skye)	1857	1962		Operational Cottages privately owned
Inchcolm (Fife)	1858			Discontinued
Rubha Nan Gall (Mull)	1857	1960		Operational Cottages privately owned
Cantick Head • (Orkney)	1858	1991		Operational Cottages privately owned
Bressay • (Shetland)	1858	1989		Operational Cottages privately owned
Ruvaal • (Islay)	1859	1983		Operational Cottages privately owned
Corran Point (Inverness-shire)	1860	1970		Operational Cottages privately owned
Fladda (Firth of Lorne)	1860	1956		Operational Cottages privately owned
McArthur's Head (Islay)	1861	1969	Cottages demolished	Operational
St Abb's Head • (Berwickshire)	1862	1993		Operational RACON Cottages privately owned
Holburn Head (Caithness)	1862	1988		Discontinued 2003
Butt of Lewis • (Lewis)	1862	1998		Operational DGPS Station AIS Station
Monach Isles • (North Uist)	1864		Discontinued 1942 Brought back in to service 2009	Operational
Skervuile (Jura)	1865	1945		Operational

Lighthouse	Established	Automated	Notable Historic Event	Status (2010)
Auskerry (Orkney)	1867	1961		Operational / Cottages privately owned
Loch Indaal (Islay)	1869	1950 Approx		Operational
Scurdie Ness (Montrose)	1870	1987		Operational RACON / Cottages privately owned
Stoer Head (Sutherland)	1870	1978		Operational / Holiday Homes
Dubh Artach (Argyll)	1872	1971	Red Band added 1890	Operational
Turnberry (Ayrshire)	1873	1986		Operational / Cottages privately owned
Chicken Rock (Isle of Man)	1875	1961	Fire 1961	Operational
Holy Island (Inner) (Arran)	1877	1977		Operational / Cottages privately owned
Langness (Isle of Man)	1880	1996		Operational / Cottages privately owned

Thomas and David A. Stevenson

Lighthouse	Established	Automated	Notable Historic Event	Status (2010)
Fidra (East Lothian)	1885	1970		Operational
Oxcars (Midlothian)	1886	1894	Transferred to Forth Ports 1990s	
Ailsa Craig (South Ayrshire)	1886	1990		Operational / Cottages privately owned

David A. Stevenson

Note: The majority of lights were built as Automatic.
The Northern Lighthouse Board's programme to remove gas from installations and convert to solar power meant the removal of

a number of minor lights. The majority where replaced by aluminium framework towers clad with GRP (Glass Reinforced Plastic) daymark boards, or sectional GRP towers to reduce the ongoing maintenance burden of the structures.

Name	Established	Rebuilt	Year	Notes	Status
Grey Rocks (Sound of Mull)	1890	Rebuilt 1991			Operational
Dubh Sgeir (Barra)	1891	Rebuilt 2003			Operational
Weaver Point (North Uist)	1891	Rebuilt 2003			Operational
Sgeir Leadh (Castlebay, Barra)	1891				Discontinued 2004
Calvay (South Uist)	1891	Rebuilt 1985			Operational
Dunollie (Oban)	1892				Operational
Crowlin (Ross & Cromarty)	1892	Rebuilt 2001			Operational
Kyle Rhea (Skye)	1892				Operational
Fair Isle South (Skaddan) (Shetland)	1892		1998	Scotland's last manned lighthouse	Operational / Cottages privately owned
Fair Isle North (Skroo) (Shetland)	1892		1983		Operational / Cottages demolished at time of Automation
Reisa an t'Struith (Jura)	1892	Rebuilt 2001			Operational
Carloway (Lewis)	1892	Rebuilt 2000			Transferred to Western Isles Council 1990s
Helliar Holm (Orkney)	1893		1967		Operational

Lighthouse	Established	Automated	Notable Historic Event	Status (2010)
Heston Island (Kirkcudbright)	1893		Rebuilt 1996	Operational
Fugla Ness (Shetland)	1893			Operational
Eyre Point (Raasay)	1893		Rebuilt 2001	Operational
Dunvegan (Skye)	1893		Rebuilt 2002	Operational
Vaila Sound (Shetland)	1894		Rebuilt 2005	Operational
Loch Eribol (Sutherland)	1894		Rebuilt 2003	Operational
Rattray Head • (Aberdeenshire)	1895	1982		Operational RACON
Skerry of Ness (Orkney)	1895		Rebuilt 1981	Operational
Hillswick (Shetland)	1895		Rebuilt 2001	Operational
Balta Sound (Shetland)	1895		Rebuilt 2003	Operational
Sule Skerry • (West of Orkney)	1895	1982		Operational RACON
Stroma • (Caithness)	1896	1996		Operational
Tod Head (Kincardineshire)	1896	1988		Discontinued 2007
Scarinish • (Tiree, Argyll)	1897			Operational
No Ness (Shetland)	1897		Discontinued	

Name (Location)	Year			Status
Muckle Roe (Shetland)	1897		Rebuilt 2001	Operational
Cava (Orkney)	1898		Rebuilt 1988	Operational
Noup Head (Westray, Orkney)	1898	1964		Operational
Flannan Isles (Ross & Cromarty)	1899	1971	Disappearance of the three keepers 1900	Operational
Tiumpan Head (Lewis)	1900	1985		Operational Cottages privately owned
Greinam (Lewis)	1900			Owned/operated by Western Isles Council
Killantringan (Wigtownshire)	1900	1988		Discontinued 2007
Duart Point (Mull)	1901		Memorial to William Black (Novelist)	Operational
Barns Ness (East Lothian)	1901	1986		Discontinued 2007
Bunessan (Mull)	1901		Rebuilt 2001	Operational
Hoxa Head (Orkney)	1901		Rebuilt 1996	Operational
Bass Rock (East Lothian)	1902	1988		Operational
Scalasaig (Colonsay)	1903		Rebuilt 2003	Operational
Lady Isle (Ayrshire)	1903			Operational RACON
The Garvellachs (Firth of Lorn)	1904		Rebuilt 2003	Operational
Suther Ness (Shetland)	1904		Rebuilt 1995	Operational

Lighthouse	Established	Automated	Notable Historic Event	Status (2010)
Whitehill (Shetland)	1904		Rebuilt 2001	Operational
Sgeir Bhuidhe (Port Appin)	1904		Rebuilt 2002	Operational
Mull of Eswick (Shetland)	1904		Fell into the sea 1994 New light established 1995	Operational
Craigton Point (Inverness)	1904		Rebuilt 2001	Operational
Symbister Ness (Shetland)	1904		Rebuilt 1997	Operational
Rova Head (Shetland)	1904		Rebuilt 1998	Operational
Hyskeir • (Inverness-shire)	1904	1997		Operational
Ardtreck (Skye)	1904		Rebuilt 2002	Operational
Roseness (Orkney)	1905		Rebuilt 1983	Operational
Holy Island (Outer) • (Arran)	1905	1977		Operational Cottages privately owned
Swona (Orkney)	1906		Rebuilt 1984	Operational
Green Island (Sound of Mull)	1906		Rebuilt 2001	Operational
Ruadh Sgeir (Jura)	1906		Rebuilt 1983	Operational
Eigg	1906		Rebuilt 1985	Operational
Lady Rock (Lismore)	1907		Rebuilt 2001	Operational
Eilean a'Chuirn (Islay)	1907			Operational

Name	Established	Rebuilt	Status
Papa Stronsay (Orkney)	1907	Rebuilt 2002	Operational
Canna	1907	Rebuilt 1986	Operational
Elie Ness • (Fife)	1908		Operational
Ornsay Beacon (Skye)	1908	Rebuilt 2002	Operational
Eilean Trodday (Skye)	1908	Rebuilt 2002	Operational
Firths Voe • (Shetland)	1909		Operational
Cairns of Coll (Coll)	1909	Rebuilt 1987	Operational
Ruff Reef (Orkney)	1909		Operational
Ness of Sound (Shetland)	1909	Rebuilt 1998	Operational
Calf of Eday (Orkney)	1909		Operational
Dubh Sgeir, Luing (Sound of Luing)	1910	Rebuilt 1990	Operational RACON
Sandaig (Glenelg)	1910	Rebuilt 2002	Operational Original tower now an information point at Glenelg Ferry
Lother Rock (Orkney)	1910		Operational RACON
Neist Point • (Skye)	1910	1990	Operational AIS Station Cottages privately owned
Na Cuiltean (Jura)	1911	Rebuilt 2004	Operational
Rubh Reidh • (Gairloch)	1912	1986	Operational AIS Station Cottages privately owned

Lighthouse	Established	Automated	Notable Historic Event	Status (2010)
Milaid Point (Lewis)	1912		Rebuilt 1998	Operational
Crammag Head • (Wigtownshire)	1913		Rebuilt 2009	Operational
Cairnbulg Briggs (Aberdeenshire)	1914			Operational
Maughold Head • (Isle of Man)	1914	1993		Operational Cottages privately owned
Copinsay • (Orkney)	1915	1991		Operational Cottages privately owned
Clythness (Caithness)	1916	1964		Discontinued 2010
Vaternish (Skye)	1924			Operational
Duncansby Head • (Caithness)	1924	1997	House demolished in 2002	Operational RACON
Brough of Birsay • (Orkney)	1925			Operational
Ardtornish (Sound of Mull)	1927		Rebuilt 2001	Operational
Carragh Mhor (Islay)	1928			Operational
Esha Ness • (Shetland)	1929	1974		Operational Cottages privately owned
Sleat Point (Skye)	1934		Rebuilt 2003	Operational
Tor Ness • (Orkney)	1937		Rebuilt 1990	Operational
Rubh Uisenis (Lewis)	1938		Rebuilt 1989	Operational

JOHN RENNIE AND ROBERT STEVENSON
BELL ROCK LIGHTHOUSE
Image reproduced with the kind permission of Ian Cowe © www.flickr.com/iancowe

The Bell Rock Lighthouse, designed and built by John Rennie and Robert Stevenson. Established 1811. The Bell Rock Lighthouse is the oldest sea-washed rock lighthouse still in continuous use.

ALAN STEVENSON
NORTH RONALDSAY LIGHTHOUSE

All images reproduced with the kind permission of Ian Cowe © www.flickr.com/iancowe

Above: North Ronaldsay Lighthouse, Orkney, designed and built by Alan Stevenson. Established 1854. This light replaced the original beacon established by Thomas Smith in 1789.

Below left: Equipment manufactured by Milne & Son, Edinburgh, 1854.

Below right: North Ronaldsay still has its original Fresnel lens.

THOMAS SMITH
PENTLAND SKERRIES LIGHTHOUSE

Top image reproduced with the kind permission of Ian Cowe © www.flickr.com/iancowe
Bottom image Northern Lighthouse Board collection

Two lighthouses, 60ft apart, were built on Pentland Skerries by Thomas Smith. Established 1794. These lights were later replaced by taller towers, 100ft apart with Robert Stevenson as the Engineer. Today only one light is operational.

BUILT 1794
THOMAS SMITH ENGINEER
ROBERT STEVENSON CLERK of WORKS
REBUILT 1830
ROBERT STEVENSON ENGINEER
ALAN STEVENSON CLERK of WORKS

ROBERT STEVENSON
TARBAT NESS LIGHTHOUSE

All images reproduced with the kind permission of Ian Cowe © www.flickr.com/iancowe

Above: Tarbat Ness Lighthouse designed and built by Robert Stevenson. Established 1830.

Left: Lantern showing the emergency back up light.

Opposite page: The view from the top of the tower. Tarbat Ness is the third tallest lighthouse in Scotland (North Ronaldsay and Skerryvore being taller) and bears two distinguishing broad red bands.

DAVID A. STEVENSON
FAIR ISLE SOUTH & NORTH LIGHTHOUSES

All images reproduced with the kind permission of Ian Cowe © www.flickr.com/iancowe

Above: Fair Isle South (Skaddan) Lighthouse was designed and built by David A. Stevenson.
Established 1892. Fair Isle South was automated in 1998 and
was Scotland's last manned lighthouse.

Below: Fair Isle North (Skroo) built by David A. Stevenson. Established 1892.
The solar panels were added in 2006.

DAVID A. STEVENSON
FLANNAN ISLES LIGHTHOUSE

All images reproduced with the kind permission of Ian Cowe © www.flickr.com/iancowe

Flannan Isles Lighthouse designed and built by David A. Stevenson. Established 1899. Most notable for the disappearance of the three keepers in December 1900.

Above: The former Shore-Station for the Flannan Isles at Breasclete.

Below: The fine detail from above the door at the Shore Station. Inscription: IN SALUTEM OMNIUM 1899

ROBERT STEVENSON
ISLE OF MAY LIGHTHOUSE

All images reproduced with the kind permission of Ian Cowe © www.flickr.com/iancowe

Isle of May Lighthouse designed and built by Robert Stevenson. Established 1816. The Isle of May was the site of the earliest coal-fired light in Scotland dating back to 1636, The original tower, though modified, is still in existence.

Internal view of the Isle of May stairwell painted to give the effect of dressed ashlar stone.

DAVID AND THOMAS STEVENSON
SCURDIE NESS LIGHTHOUSE

All images reproduced with the kind permission of Ian Cowe © www.flickr.com/iancowe

Above: Scurdie Ness Lighthouse was designed and built by David and Thomas Stevenson. Established 1870.

Right: The internal brick tube from which the original clockwork weight would rise and fall, now used to channel the cables connected to the automation of the station.

ALAN STEVENSON
ARDNAMURCHAN LIGHTHOUSE

Top image reproduced with the kind permission of the Northern Lighthouse Board: Photograph by
Arnaud Späni ©. Bottom: Northern Lighthouse Board Collection

Above: Ardnamurchan Lighthouse was designed and built by Alan Stevenson. Established 1849. The light stands on the most westerly point on the Scottish mainland.

Left: Egyptian influences can be seen in the ornate corbel beneath the balcony and doorway.

ALAN STEVENSON
SKERRYVORE LIGHTHOUSE

All images reproduced with the kind permission of Ian Cowe © www.flickr.com/iancowe

Skerryvore Lighthouse was designed and built by Alan Stevenson. Established 1844. It stands 48 metres high and is built of pink Ross of Mull granite, apart from the four lower courses of gneiss rock from Tiree.

Above: Skerryvore Lighthouse – base of tower (see page 12)

Below: Hynish was chosen as the location for a shore station and harbour as it was the closest location to Skerryvore, the harbour regularly silted up and Alan Stevenson had to design a fresh water flushing system connected to a reservoir to remove the silt.

ROBERT STEVENSON
CORSEWALL LIGHTHOUSE

All images reproduced with the kind permission of the Northern Lighthouse Board: Photographs by Arnaud Späni ©.

Above: Corsewall Lighthouse designed and built by Robert Stevenson. Established 1817.

Below left: Corsewall is one of the few Stevenson lights to have an open spiral staircase.

Below right: Following the automation of the station in 1994 the former keepers' cottages were sold and converted into the Corsewall Lighthouse Hotel.

DAVID A. STEVENSON
RATTRAY HEAD LIGHTHOUSE
Image reproduced with the kind permission of Ian Cowe © www.flickr.com/iancowe

Rattray Head Lighthouse designed and built by David A. Stevenson. Established 1895. The lower section is built of granite blocks, mostly quarried from Rubislaw, and formerly housed the fog horn and engine room. The upper section is built of brick.

ROBERT STEVENSON
GIRDLE NESS LIGHTHOUSE

All images reproduced with the kind permission of Ian Cowe © www.flickr.com/iancowe

Girdle Ness Lighthouse designed and built by Robert Stevenson. Established 1833. Originally the light showed two lights, both fixed. The lower light was in a glazed gallery about one third of the way up the tower. The lower light was discontinued in 1890.

Right: The fine detail from the lantern astragal joint.

Below: The sealed beam lamp array installed at automation.

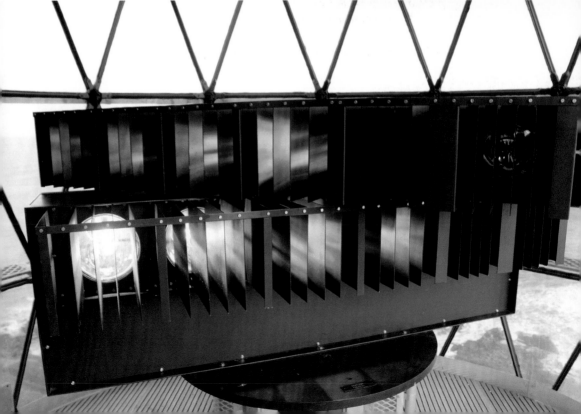

ALAN STEVENSON
COVESEA SKERRIES LIGHTHOUSE

All images reproduced with the kind permission of Ian Cowe © www.flickr.com/iancowe

Above: Covesea Skerries Lighthouse designed and built by Alan Stevenson. Established 1846.
Egyptian influences can be seen in the ornate corbel beneath the balcony.
In the foreground is the original sundial pedestal.

Right: The detail from the canopy above the door to the tower.

DAVID A. STEVENSON
NEIST POINT LIGHTHOUSE

All images reproduced with the kind permission of Ian Cowe © www.flickr.com/iancowe

Neist Point Lighthouse designed and built by David A. Stevenson. Established 1910. Mr W. Hugh MacDonald from Oban was the building contractor for the lighthouse and dwellings which cost £4,350.

Index of Lighthouses and Lights

Note: figures in bold type indicate an illustration

Dynasty of Engineers — General Index

Note: figures in bold type indicate an illustration